Pro Microsoft HDInsight

Hadoop on Windows

Debarchan Sarkar

Apress

Pro Microsoft HDInsight: Hadoop on Windows

ISBN-13 (pbk): 978-1-4302-6055-4

ISBN-13 (electronic): 978-1-4302-6056-1

President and Publisher: Paul Manning
Lead Editor: Jonathan Gennick
Technical Reviewer: Scott Klein, Rodney Landrum
Editorial Board: Steve Anglin, Mark Beckner, Ewan Buckingham, Gary Cornell, Louise Corrigan, James T. DeWolf, Jonathan Gennick, Jonathan Hassell, Robert Hutchinson, Michelle Lowman, James Markham, Matthew Moodie, Jeff Olson, Jeffrey Pepper, Douglas Pundick, Ben Renow-Clarke, Dominic Shakeshaft, Gwenan Spearing, Matt Wade, Steve Weiss
Coordinating Editor: Anamika Panchoo
Copy Editor: Roger LeBlanc
Compositor: SPi Global
Indexer: SPi Global
Artist: SPi Global
Cover Designer: Anna Ishchenko

Distributed to the book trade worldwide by Springer Science+Business Media New York, 233 Spring Street, 6th Floor, New York, NY 10013. Phone 1-800-SPRINGER, fax (201) 348-4505, e-mail orders-ny@springer-sbm.com, or visit www.springeronline.com. Apress Media, LLC is a California LLC and the sole member (owner) is Springer Science + Business Media Finance Inc (SSBM Finance Inc). SSBM Finance Inc is a Delaware corporation.

For information on translations, please e-mail rights@apress.com, or visit www.apress.com.

Apress and friends of ED books may be purchased in bulk for academic, corporate, or promotional use. eBook versions and licenses are also available for most titles. For more information, reference our Special Bulk Sales–eBook Licensing web page at www.apress.com/bulk-sales.

Any source code or other supplementary material referenced by the author in this text is available to readers at www.apress.com. For detailed information about how to locate your book's source code, go to www.apress.com/source-code/.

I dedicate my work to my mother, Devjani Sarkar.

All that I am, or hope to be, I owe to you my Angel Mother. You have been my inspiration throughout my life. I learned commitment, responsibility, integrity and all other values of life from you. You taught me everything, to be strong and focused, to fight honestly against every hardship in life. I know that I could not be the best son, but trust me, each day when I wake up, I think of you and try to spend the rest of my day to do anything and everything just to see you more happy and proud to be my mother. Honestly, I never even dreamed of publishing a book some day. Your love and encouragement have been the fuel that enabled me to do the impossible. You've been the bones of my spine, keeping me straight and true. You're my blood, making sure it runs rich and strong. You're the beating of my heart. I cannot imagine a life without you, Love you so much MA!

Contents at a Glance

Contents

About the Author

Debarchan Sarkar (@debarchans) is a Senior Support Engineer on the Microsoft HDInsight team and a technical author of books on SQL Server BI and Big Data. His total tenure at Microsoft is 6 years, and he was with SQL Server BI team before diving deep into Big Data and the Hadoop world. He is an SME in SQL Server Integration Services and is passionate about the present-day Microsoft self-service BI tools and data analysis, especially social-media brand sentiment analysis. Debarchan hails from the "city of joy," Calcutta, India and is presently located in Bangalore, India for his job in Microsoft's Global Technical Support Center. Apart from his passion for technology, he is interested in visiting new places, listening to music—the greatest creation ever on Earth—meeting new people, and learning new things because he is a firm believer that "Known is a drop; the unknown is an ocean." On a lighter note, he thinks it's pretty funny when people talk about themselves in the third person.

About the Technical Reviewers

Rodney Landrum went to school to be a poet and a writer. And then he graduated, so that dream was crushed. He followed another path, which was to become a professional in the fun-filled world of Information Technology. He has worked as a systems engineer, UNIX and network admin, data analyst, client services director, and finally as a database administrator. The old hankering to put words on paper, while paper still existed, got the best of him, and in 2000 he began writing technical articles, some creative and humorous, some quite the opposite. In 2010, he wrote *The SQL Server Tacklebox*, a title his editor disdained, but a book closest to the true creative potential he sought; he wanted to do a full book without a single screen shot. He promises his next book will be fiction or a collection of poetry, but that has yet to transpire.

Scott Klein is a Microsoft Data Platform Technical Evangelist who lives and breathes data. His passion for data technologies brought him to Microsoft in 2011 which has allowed him to travel all over the globe evangelizing SQL Server and Microsoft's cloud data services. Prior to Microsoft Scott was one of the first 4 SQL Azure MVPs, and even though those don't exist anymore, he still claims it. Scott has authored several books that talk about SQL Server and Windows Azure SQL Database and continues to look for ways to help people and companies grok the benefits of cloud computing. He also thinks "grok" is an awesome word. In his spare time (what little he has), Scott enjoys spending time with his family, trying to learn German, and has decided to learn how to brew root beer (without using the extract). He recently learned that data scientists are "sexy" so he may have to add that skill to his toolbelt.

Acknowledgments

This book benefited from a large and wide variety of people, ideas, input, and efforts. I'd like to acknowledge several of them and apologize in advance to those I may have forgotten—I hope you guys will understand.

My heartfelt and biggest THANKS perhaps, is to *Andy Leonard (@AndyLeonard)* for his help on this book project. Without Andy, this book wouldn't have been a reality. Thanks Andy, for trusting me and making it possible for me to realize my dream. I truly appreciate the great work you and *Linchpin People* are doing for the SQL Server and BI community, helping SQL Server to be a better product each day.

Thanks to the folks at Apress, *Ana* and *Jonathan* for their patience; *Roger* for his excellent, accurate, and insightful copy editing; and *Rodney* and *Scott* for their supportive comments and suggestions during the author reviews.

I would also like to thank two of my colleagues: *Krishnakumar Rukmangathan* for helping me with some of the diagrams for the book, and *Amarpreet Singh Bassan* for his help in authoring the chapters on troubleshooting. You guys were of great help. Without your input, it would have been a struggle and the book would have been incomplete.

Last but not least, I must acknowledge all the support and encouragement provided by my good friends *Sneha Deep Chowdhury* and *Soumendu Mukherjee.* Though you are experts in completely different technical domains, you guys have always been there with me listening patiently about the progress of the book, the hurdles faced and what not, from the beginning to the end. Thanks for being there with me through all my blabberings.

Introduction

My journey in Big Data started back in 2012 in one of our unit meetings. Ranjan Bhattacharjee (our boss) threw in some food for thought with his questions: "Do you guys know Big Data? What do you think about it?" That was the first time I heard the phrase "Big Data." His inspirational speech on Big Data, Hadoop, and future trends in the industry, triggered the passion for learning something new in a few of us.

Now we are seeing results from a historic collaboration between open source and proprietary products in the form of Microsoft HDInsight. Microsoft and Apache have joined hands in an effort to make Hadoop available on Windows, and HDInsight is the result. I am a big fan of such integration. I strongly believe that the future of IT will be seen in the form of integration and collaboration opening up new dimensions in the industry.

The world of data has seen exponential growth in volume in the past couple of years. With the web integrated in each and every type of device, we are generating more digital data every two years than the volume of data generated since the dawn of civilization. Learning the techniques to store, manage, process, and most importantly, make sense of data is going to be key in the coming decade of data explosion. Apache Hadoop is already a leader as a Big Data solution framework based on Java/Linux. This book is intended for readers who want to get familiar with HDInsight, which is Microsoft's implementation of Apache Hadoop on Windows.

Microsoft HDInsight is currently available as an Azure service. Windows Azure HDInsight Service brings in the user friendliness and ease of Windows through its blend of Infrastructure as a Service (IaaS) and Platform as a Service (PaaS). Additionally, it introduces .NET and PowerShell based job creation, submission, and monitoring frameworks for the developer communities based on Microsoft platforms.

Intended Audience

Pro Microsoft HDInsight is intended for people who are already familiar with Apache Hadoop and its ecosystem of projects. Readers are expected to have a basic understanding of Big Data as well as some working knowledge of present-day Business Intelligence (BI) tools. This book specifically covers HDInsight, which is Microsoft's implementation of Hadoop on Windows. The book covers HDInsight and its tight integration with the ecosystem of other Microsoft products, like SQL Server, Excel, and various BI tools. Readers should have some understanding of those tools in order to get the most from this book.

Versions Used

It is important to understand that HDInsight is offered as an Azure service. The upgrades are pretty frequent and come in the form of Azure Service Updates. Additionally, HDInsight as a product has core dependencies on Apache Hadoop. Every change in the Apache project needs to be ported as well. Thus, you should expect that version numbers of several components will be updated and changed going forward. However, the crux of Hadoop and HDInsight is not going to change much. In other words, the core of this book's content and methodologies are going to hold up well.

Structure of the Book

This book is best read sequentially from the beginning to the end. I have made an effort to provide the background of Microsoft's Big Data story, HDInsight as a technology, and the Windows Azure Storage infrastructure. This book gradually takes you through a tour of HDInsight cluster creation, job submission, and monitoring, and finally ends with some troubleshooting steps.

Chapter 1 – "Introducing HDInsight" starts off the book by giving you some background on Big Data and the current market trends. This chapter has a brief overview of Apache Hadoop and its ecosystem and focuses on how HDInsight evolved as a product.

Chapter 2 – "Understanding Windows Azure HDInsight Service" introduces you to Microsoft's Azure-based service for Apache Hadoop. This chapter discusses the Azure HDInsight service and the underlying Azure storage infrastructure it uses. This is a notable difference in Microsoft's implementation of Hadoop on Windows Azure, because it isolates the storage and the cluster as a part of the elastic service offering. Running idle clusters only for storage purposes is no longer the reality, because with the Azure HDInsight service, you can spin up your clusters only during job submission and delete them once the jobs are done, with all your data safely retained in Azure storage.

Chapter 3 – "Provisioning Your HDInsight Service Cluster" takes you through the process of creating your Hadoop clusters on Windows Azure virtual machines. This chapter covers the Windows Azure Management portal, which offers you step-by-step wizards to manually provision your HDInsight clusters in a matter of a few clicks.

Chapter 4 – "Automating HDInsight Cluster Provisioning" introduces the Hadoop .NET SDK and Windows PowerShell cmdlets to automate cluster-creation operations. Automation is a common need for any business process. This chapter enables you to create such configurable and automatic cluster-provisioning based on C# code and PowerShell scripts.

Chapter 5 – "Submitting Jobs to Your HDInsight Cluster" shows you ways to submit MapReduce jobs to your HDInsight cluster. You can leverage the same .NET and PowerShell based framework to submit your data processing operations and retrieve the output. This chapter also teaches you how to create a MapReduce job in .NET. Again, this is unique in HDInsight, as traditional Hadoop jobs are based on Java only.

Chapter 6 – "Exploring the HDInsight Name Node" discusses the Azure virtual machine that acts as your cluster's Name Node when you create a cluster. You can log in remotely to the Name Node and execute command-based Hadoop jobs manually. This chapter also speaks about the web applications that are available by default to monitor cluster health and job status when you install Hadoop.

Chapter 7 – "Using the Windows Azure HDInsight Emulator" introduces you to the local, one-box emulator for your Azure service. This emulator is primarily intended to be a test bed for testing or evaluating the product and your solution before you actually roll it out to Azure. You can simulate both the HDInsight cluster and Azure storage so that you can evaluate it absolutely free of cost. This chapter teaches you how to install the emulator, set the configuration options, and test run MapReduce jobs on it using the same techniques.

Chapter 8 – "Accessing HDInsight over Hive and ODBC" talks about the ODBC endpoint that the HDInsight service exposes for client applications. Once you install and configure the ODBC driver correctly, you can consume the Hive service running on HDInsight from any ODBC-compliant client application. This chapter takes you through the download, installation, and configuration of the driver to the successful connection to HDInsight.

Chapter 9 – "Consuming HDInsight from Self-Service BI Tools" is a particularly interesting chapter for readers who have a BI background. This chapter introduces some of the present-day, self-service BI tools that can be set up with HDInsight within a few clicks. With data visualization being the end goal of any data-processing framework, this chapter gets you going with creating interactive reports in just a few minutes.

Chapter 10 – "Integrating HDInsight with SQL Server Integration Services" covers the integration of HDInsight with SQL Server Integration Services (SSIS). SSIS is a component of the SQL Server BI suite and plays an important part in data-processing engines as a data extract, transform, and load tool. This chapter guides you through creating an SSIS package that moves data from Hive to SQL Server

Chapter 11 – "Logging in HDInsight" describes the logging mechanism in HDInsight. There is built-in logging in Apache Hadoop; on top of that, HDInsight implements its own logging framework. This chapter enables readers to learn about the log files for the different services and where to look if something goes wrong.

Chapter 12 – "Troubleshooting Cluster Deployments" is about troubleshooting scenarios you might encounter during your cluster-creation process. This chapter explains the different stages of a cluster deployment and the deployment logs on the Name Node, as well as offering some tips on troubleshooting C# and PowerShell based deployment scripts.

Chapter 13 – "Troubleshooting Job Failures" explains the different ways of troubleshooting a MapReduce job-execution failure. This chapter also speaks about troubleshooting performance issues you might encounter, such as when jobs are timing out, running out of memory, or running for too long. It also covers some best-practice scenarios.

Downloading the Code

The author provides code to go along with the examples in this book. You can download that example code from the book's catalog page on the Apress.com website. The URL to visit is http://www.apress.com/9781430260554. Scroll about halfway down the page. Then find and click the tab labeled Source Code/Downloads.

Contacting the Author

You can contact the author, Debarchan Sarkar, through his twitter handle *@debarchans*. You can also follow his Facebook group at https://www.facebook.com/groups/bigdatalearnings/ and his Facebook page on HDInsight at https://www.facebook.com/MicrosoftBigData.

CHAPTER 1

■ ■ ■

Introducing HDInsight

HDInsight is Microsoft's distribution of "Hadoop on Windows." Microsoft has embraced Apache Hadoop to provide business insight to all users interested in tuning raw data into meaning by analyzing all types of data, structured or unstructured, of any size. The new Hadoop-based distribution for Windows offers IT professionals ease of use by simplifying the acquisition, installation and configuration experience of Hadoop and its ecosystem of supporting projects in Windows environment. Thanks to smart packaging of Hadoop and its toolset, customers can install and deploy Hadoop in hours instead of days using the user-friendly and flexible cluster deployment wizards.

This new Hadoop-based distribution from Microsoft enables customers to derive business insights on structured and unstructured data of any size and activate new types of data. Rich insights derived by analyzing Hadoop data can be combined seamlessly with the powerful Microsoft Business Intelligence Platform. The rest of this chapter will focus on the current data-mining trends in the industry, the limitations of modern-day data-processing technologies, and the evolution of HDInsight as a product.

What Is Big Data, and Why Now?

All of a sudden, everyone has money for Big Data. From small start-ups to mid-sized companies and large enterprises, businesses are now keen to invest in and build Big Data solutions to generate more intelligent data. So what is Big Data all about?

In my opinion, *Big Data* is the new buzzword for a data mining technology that has been around for quite some time. Data analysts and business managers are fast adopting techniques like predictive analysis, recommendation service, clickstream analysis etc. that were commonly at the core of data processing in the past, but which have been ignored or lost in the rush to implement modern relational database systems and structured data storage. Big Data encompasses a range of technologies and techniques that allow you to extract useful and previously hidden information from large quantities of data that previously might have been left dormant and, ultimately, thrown away because storage for it was too costly.

Big Data solutions aim to provide data storage and querying functionality for situations that are, for various reasons, beyond the capabilities of traditional database systems. For example, analyzing social media sentiments for a brand has become a key parameter for judging a brand's success. Big Data solutions provide a mechanism for organizations to extract meaningful, useful, and often vital information from the vast stores of data that they are collecting.

Big Data is often described as a solution to the "three V's problem":

> **Variety:** It's common for 85 percent of your new data to not match any existing data schema. Not only that, it might very well also be semi-structured or even unstructured data. This means that applying schemas to the data before or during storage is no longer a practical option.

> **Volume:** Big Data solutions typically store and query thousands of terabytes of data, and the total volume of data is probably growing by ten times every five years. Storage solutions must be able to manage this volume, be easily expandable, and work efficiently across distributed systems.

Velocity: Data is collected from many new types of devices, from a growing number of users and an increasing number of devices and applications per user. Data is also emitted at a high rate from certain modern devices and gadgets. The design and implementation of storage and processing must happen quickly and efficiently.

Figure 1-1 gives you a theoretical representation of Big Data, and it lists some possible components or types of data that can be integrated together.

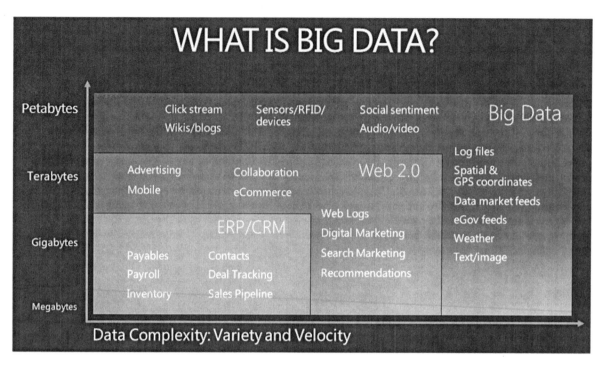

Figure 1-1. *Examples of Big Data and Big Data relationships*

There is a striking difference in the ratio between *the speeds at which data is generated compared to the speed at which it is consumed* in today's world, and it has always been like this. For example, today a standard international flight generates around .5 terabytes of operational data. That is during a single flight! Big Data solutions were already implemented long ago, back when Google/Yahoo/Bing search engines were developed, but these solutions were limited to large enterprises because of the hardware cost of supporting such solutions. This is no longer an issue because hardware and storage costs are dropping drastically like never before. New types of questions are being asked and data solutions are used to answer these questions and drive businesses more successfully. These questions fall into the following categories:

- **Questions regarding social and Web analytics:** Examples of these types of questions include the following: What is the sentiment toward our brand and products? How effective are our advertisements and online campaigns? Which gender, age group, and other demographics are we trying to reach? How can we optimize our message, broaden our customer base, or target the correct audience?

- **Questions that require connecting to live data feeds:** Examples of this include the following: a large shipping company that uses live weather feeds and traffic patterns to fine-tune its ship and truck routes to improve delivery times and generate cost savings; retailers that analyze sales, pricing, economic, demographic, and live weather data to tailor product selections at particular stores and determine the timing of price markdowns.

- **Questions that require advanced analytics:** An examples of this type is a credit card system that uses machine learning to build better fraud-detection algorithms. The goal is to go beyond the simple business rules involving charge frequency and location to also include an individual's customized buying patterns, ultimately leading to a better experience for the customer.

Organizations that take advantage of Big Data to ask and answer these questions will more effectively derive new value for the business, whether it is in the form of revenue growth, cost savings, or entirely new business models. One of the most obvious questions that then comes up is this: What is the shape of Big Data?

Big Data typically consists of delimited attributes in files (for example, comma separated value, or CSV format), or it might contain long text (tweets), Extensible Markup Language (XML),Javascript Object Notation (JSON)and other forms of content from which you want only a few attributes at any given time. These new requirements challenge traditional data-management technologies and call for a new approach to enable organizations to effectively manage data, enrich data, and gain insights from it.

Through the rest of this book, we will talk about how Microsoft offers an end-to-end platform for all data, and the easiest to use tools to analyze it. Microsoft's data platform seamlessly manages any data (relational, nonrelational and streaming) of any size (gigabytes, terabytes, or petabytes) anywhere (on premises and in the cloud), and it enriches existing data sets by connecting to the world's data and enables all users to gain insights with familiar and easy to use tools through Office, SQL Server and SharePoint.

How Is Big Data Different?

Before proceeding, you need to understand the difference between traditional relational database management systems (RDBMS) and Big Data solutions, particularly how they work and what result is expected.

Modern relational databases are highly optimized for fast and efficient query processing using different techniques. Generating reports using Structured Query Language (SQL) is one of the most commonly used techniques.

Big Data solutions are optimized for reliable storage of vast quantities of data; the often unstructured nature of the data, the lack of predefined schemas, and the distributed nature of the storage usually preclude any optimization for query performance. Unlike SQL queries, which can use indexes and other intelligent optimization techniques to maximize query performance, Big Data queries typically require an operation similar to a full table scan. Big Data queries are batch operations that are expected to take some time to execute.

You can perform real-time queries in Big Data systems, but typically you will run a query and store the results for use within your existing business intelligence (BI) tools and analytics systems. Therefore, Big Data queries are typically batch operations that, depending on the data volume and query complexity, might take considerable time to return a final result. However, when you consider the volumes of data that Big Data solutions can handle, which are well beyond the capabilities of traditional data storage systems, the fact that queries run as multiple tasks on distributed servers does offer a level of performance that cannot be achieved by other methods. Unlike most SQL queries used with relational databases, Big Data queries are typically not executed repeatedly as part of an application's execution, so batch operation is not a major disadvantage.

Is Big Data the Right Solution for You?

There is a lot of debate currently about relational vs. nonrelational technologies. "Should I use relational or non-relational technologies for my application requirements?" is the wrong question. Both technologies are storage mechanisms designed to meet very different needs. Big Data is not here to replace any of the existing relational-model-based data storage or mining engines; rather, it will be complementary to these traditional systems, enabling people to combine the power of the two and take data analytics to new heights.

The first question to be asked here is, "Do I even need Big Data?" Social media analytics have produced great insights about what consumers think about your product. For example, Microsoft can analyze Facebook posts or Twitter sentiments to determine how Windows 8.1, its latest operating system, has been accepted in the industry and the community. Big Data solutions can parse huge unstructured data sources—such as posts, feeds, tweets, logs, and so forth—and generate intelligent analytics so that businesses can make better decisions and correct predictions. Figure 1-2 summarizes the thought process.

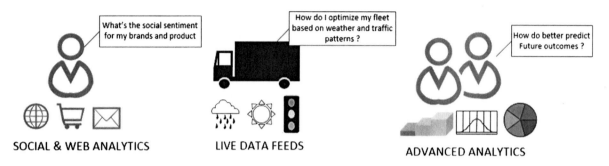

SOCIAL & WEB ANALYTICS   LIVE DATA FEEDS   ADVANCED ANALYTICS

Figure 1-2. *A process for determining whether you need Big Data*

The next step in evaluating an implementation of any business process is to know your existing infrastructure and capabilities well. Traditional RDBMS solutions are still able to handle most of your requirements. For example, Microsoft SQL Server can handle 10s of TBs, whereas Parallel Data Warehouse (PDW) solutions can scale up to 100s of TBs of data.

If you have highly relational data stored in a structured way, you likely don't need Big Data. However, both SQL Server and PDW appliances are not good at analyzing streaming text or dealing with large numbers of attributes or JSON. Also, typical Big Data solutions use a scale-out model (distributed computing) rather than a scale-up model (increasing computing and hardware resources for a single server) targeted by traditional RDBMS like SQL Server.

With hardware and storage costs falling drastically, distributed computing is rapidly becoming the preferred choice for the IT industry, which uses massive amounts of commodity systems to perform the workload.

However, to what type of implementation you need, you must evaluate several factors related to the three Vs mentioned earlier:

- **Do you want to integrate diverse, heterogeneous sources? (Variety):** If your answer to this is yes, is your data predominantly semistructured or unstructured/nonrelational data? Big Data could be an optimum solution for textual discovery, categorization, and predictive analysis.

- **What are the quantitative and qualitative analyses of the data? (Volume):** Is there a huge volume of data to be referenced? Is data emitted in streams or in batches? Big Data solutions are ideal for scenarios where massive amounts of data needs to be either streamed or batch processed.

- **What is the speed at which the data arrives? (Velocity):** Do you need to process data that is emitted at an extremely fast rate? Examples here include data from devices, radio-frequency identification device (RFID) transmitting digital data every micro second, or other such scenarios. Traditionally, Big Data solutions are batch-processing or stream-processing systems best suited for such streaming of data. Big Data is also an optimum solution for processing historic data and performing trend analyses.

Finally, if you decide you need a Big Data solution, the next step is to evaluate and choose a platform. There are several you can choose from, some of which are available as cloud services and some that you run on your own on-premises or hosted hardware. This book focuses on Microsoft's Big Data solution, which is the Windows Azure HDInsight Service. This book also covers the Windows Azure HDInsight Emulator, which provides a test bed for use before you deploy your solution to the Azure service.

The Apache Hadoop Ecosystem

The Apache open source project Hadoop is the traditional and, undoubtedly, most well-accepted Big Data solution in the industry. Originally developed largely by Google and Yahoo, Hadoop is the most scalable, reliable, distributed-computing framework available. It's based on Unix/Linux and leverages commodity hardware.

A typical Hadoop cluster might have 20,000 nodes. Maintaining such an infrastructure is difficult both from a management point of view and a financial one. Initially, only large IT enterprises like Yahoo, Google, and Microsoft could afford to invest in such Big Data solutions, such as Google search, Bing maps, and so forth. Currently, however, hardware and storage costs are going so down. This enables small companies or even consumers to think about using a Big Data solution. Because this book covers Microsoft HDInsight, which is based on core Hadoop, we will first give you a quick look at the Hadoop core components and few of its supporting projects.

The core of Hadoop is its storage system and its distributed computing model. This model includes the following technologies and features:

- **HDFS:** Hadoop Distributed File System is responsible for storing data on the cluster. Data is split into blocks and distributed across multiple nodes in the cluster.

- **MapReduce:** A distributed computing model used to process data in the Hadoop cluster that consists of two phases: Map and Reduce. Between Map and Reduce, shuffle and sort occur.

MapReduce guarantees that the input to every reducer is sorted by key. The process by which the system performs the sort and transfers the map outputs to the reducers as inputs is known as the *shuffle*. The shuffle is the heart of MapReduce, and it's where the "magic" happens. The shuffle is an area of the MapReduce logic where optimizations are made. By default, Hadoop uses Quicksort; afterward, the sorted intermediate outputs get merged together. Quicksort checks the recursion depth and gives up when it is too deep. If this is the case, Heapsort is used. You can customize the sorting method by changing the algorithm used via the `map.sort.class` value in the `hadoop-default.xml` file.

The Hadoop cluster, once successfully configured on a system, has the following basic components:

- **Name Node:** This is also called the Head Node of the cluster. Primarily, it holds the metadata for HDFS. That is, during processing of data, which is distributed across the nodes, the Name Node keeps track of each HDFS data block in the nodes. The Name Node is also responsible for maintaining heartbeat co-ordination with the data nodes to identify dead nodes, decommissioning nodes and nodes in safe mode. The Name Node is the single point of failure in a Hadoop cluster.

- **Data Node:** Stores actual HDFS data blocks. The data blocks are replicated on multiple nodes to provide fault-tolerant and high-availability solutions.

- **Job Tracker:** Manages MapReduce jobs, and distributes individual tasks.

- **Task Tracker:** Instantiates and monitors individual Map and Reduce tasks.

Additionally, there are a number of supporting projects for Hadoop, each having its unique purpose—for example, to feed input data to the Hadoop system, to be a data-warehousing system for ad-hoc queries on top of Hadoop, and many more. Here are a few specific examples worth mentioning:

- **Hive:** A supporting project for the main Apache Hadoop project. It is an abstraction on top of MapReduce that allows users to query the data without developing MapReduce applications. It provides the user with a SQL-like query language called *Hive Query Language (HQL)* to fetch data from the Hive store.

- **PIG:** An alternative abstraction of MapReduce that uses a data flow scripting language called PigLatin.

- **Flume:** Provides a mechanism to import data into HDFS as data is generated.

- **Sqoop:** Provides a mechanism to import and export data to and from relational database tables and HDFS.

- **Oozie:** Allows you to create a workflow for MapReduce jobs.

- **HBase:** Hadoop database, a *NoSQL* database.

- **Mahout:** A machine-learning library containing algorithms for clustering and classification.

- **Ambari:** A project for monitoring cluster health statistics and instrumentation.

Figure 1-3 gives you an architectural view of the Apache Hadoop ecosystem. We will explore some of the components in the subsequent chapters of this book, but for a complete reference, visit the Apache web site at `http://hadoop.apache.org/`.

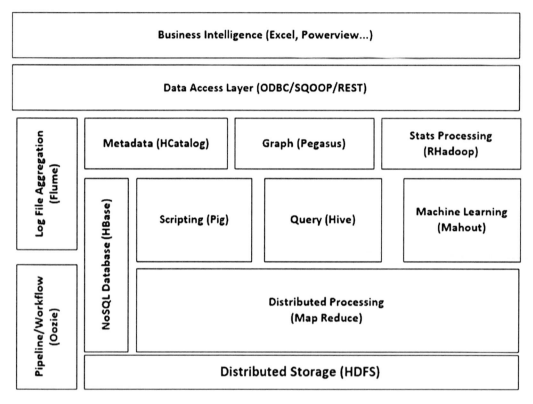

Figure 1-3. *The Hadoop ecosystem*

As you can see, deploying a Hadoop solution requires setup and management of a complex ecosystem of frameworks (often referred to as a *zoo*) across clusters of computers. This might be the only drawback of the Apache Hadoop framework—the complexity and efforts involved in creating an efficient cluster configuration and the ongoing administration required. With storage being a commodity, people are looking for easy "off the shelf" offerings for Hadoop solutions. This has led to companies like Cloudera, Green Plum and others offering their own distribution of Hadoop solutions as an out-of-the-box package. The objective is to make Hadoop solutions easily configurable as well as to make it available on diverse platforms. This has been a grand success in this era of predictive analysis through Twitter, pervasive use of social media, and the popularity of the self-service BI concept. The future of IT is integration; it could be integration between closed and open source projects, integration between unstructured and structured data, or some other form of integration. With the luxury of being able to store any type of data inexpensively, the world is looking forward to entire new dimensions of data processing and analytics.

Note HDInsight currently supports Hive, Pig, Oozie, Sqoop, and HCatalog out of the box. The plan is to also ship HBase and Flume in future versions. The beauty of HDInsight (or any other distribution) is that it is implemented on top of the Hadoop core. So you can install and configure any of these supporting projects on the default install. There is also every possibility that HDInsight will support more of these projects going forward, depending on user demand.

Microsoft HDInsight: Hadoop on Windows

HDInsight is Microsoft's implementation of a Big Data solution with Apache Hadoop at its core. HDInsight is 100 percent compatible with Apache Hadoop and is built on open source components in conjunction with Hortonworks, a company focused toward getting Hadoop adopted on the Windows platform. Basically, Microsoft has taken the open source Hadoop project, added the functionalities needed to make it compatible with Windows (because Hadoop is based on Linux), and submitted the project back to the community. All of the components are retested in typical scenarios to ensure that they work together correctly and that there are no versioning or compatibility issues.

I'm a great fan of such integration because I can see the boost it might provide to the industry, and I was excited with the news that the open source community has included Windows-compatible Hadoop in their main project trunk. Developments in HDInsight are regularly fed back to the community through Hortonworks so that they can maintain compatibility and contribute to the fantastic open source effort.

Microsoft's Hadoop-based distribution brings the robustness, manageability, and simplicity of Windows to the Hadoop environment. The focus is on hardening security through integration with Active Directory, thus making it enterprise ready, simplifying manageability through integration with System Center 2012, and dramatically reducing the time required to set up and deploy via simplified packaging and configuration.

These improvements will enable IT to apply consistent security policies across Hadoop clusters and manage them from a single pane of glass on System Center 2012. Further, Microsoft SQL Server and its powerful BI suite can be leveraged to apply analytics and generate interactive business intelligence reports, all under the same roof. For the Hadoop-based service on Windows Azure, Microsoft has further lowered the barrier to deployment by enabling the seamless setup and configuration of Hadoop clusters through an easy-to-use, web-based portal and offering Infrastructure as a Service (IaaS). Microsoft is currently the only company offering scalable Big Data solutions in the cloud and for on-premises use. These solutions are all built on a common Microsoft Data Platform with familiar and powerful BI tools.

HDInsight is available in two flavors that will be covered in subsequent chapters of this book:

- **Windows Azure HDInsight Service:** This is a service available to Windows Azure subscribers that uses Windows Azure clusters and integrates with Windows Azure storage. An Open Database Connectivity (ODBC) driver is available to connect the output from HDInsight queries to data analysis tools.

- **Windows Azure HDInsight Emulator:** This is a single-node, single-box product that you can install on Windows Server 2012, or in your Hyper-V virtual machines. The purpose of the emulator is to provide a development environment for use in testing and evaluating your solution before deploying it to the cloud. You save money by not paying for Azure hosting until after your solution is developed and tested and ready to run. The emulator is available for free and will continue to be a single-node offering.

While keeping all these details about Big Data and Hadoop in mind, it would be incorrect to think that HDInsight is a stand-alone solution or a separate solution of its own. HDInsight is, in fact, a component of the Microsoft Data Platform and part of the company's overall data acquisition, management, and visualization strategy.

Figure 1-4 shows the bigger picture, with applications, services, tools, and frameworks that work together and allow you to capture data, store it, and visualize the information it contains. Figure 1-4 also shows where HDInsight fits into the Microsoft Data Platform.

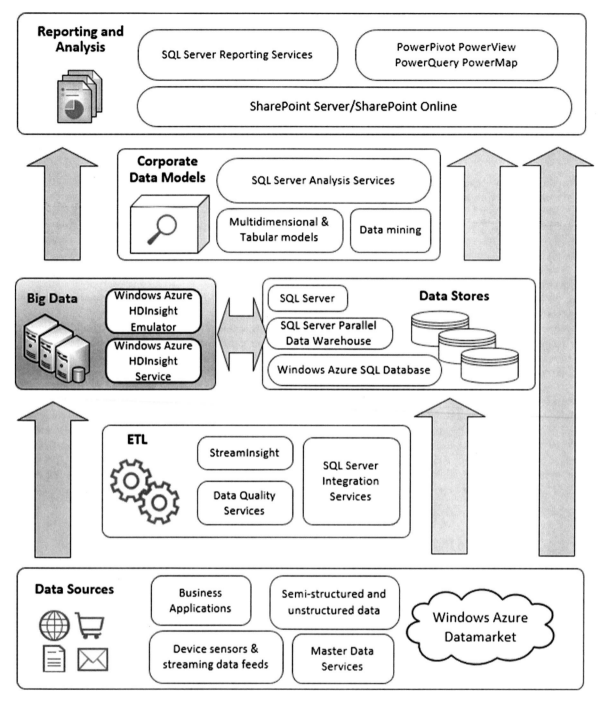

Figure 1-4. The Microsoft data platform

Combining HDInsight with Your Business Processes

Big Data solutions open up new opportunities for turning data into meaningful information. They can also be used to extend existing information systems to provide additional insights through analytics and data visualization. Every organization is different, so there is no definitive list of ways you can use HDInsight as part of your own business processes. However, there are four general architectural models. Understanding these will help you start making decisions about how best to integrate HDInsight with your organization, as well as with your existing BI systems and tools. The four different models are

- **A data collection, analysis, and visualization tool:** This model is typically chosen for handling data you cannot process using existing systems. For example, you might want to analyze sentiments about your products or services from micro-blogging sites like Twitter, social media like Facebook, feedback from customers through email, web pages, and so forth. You might be able to combine this information with other data, such as demographic data that indicates population density and other characteristics in each city where your products are sold.

- **A data-transfer, data-cleansing, and ETL mechanism:** HDInsight can be used to extract and transform data before you load it into your existing databases or data-visualization tools. HDInsight solutions are well suited to performing categorization and normalization of data, and for extracting summary results to remove duplication and redundancy. This is typically referred to as an Extract, Transform, and Load (ETL) process.

- **A basic data warehouse or commodity-storage mechanism:** You can use HDInsight to store both the source data and the results of queries executed over this data. You can also store schemas (or, to be precise, metadata) for tables that are populated by the queries you execute. These tables can be indexed, although there is no formal mechanism for managing key-based relationships between them. However, you can create data repositories that are robust and reasonably low cost to maintain, which is especially useful if you need to store and manage huge volumes of data.

- **An integration with an enterprise data warehouse and BI system:** Enterprise-level data warehouses have some special characteristics that differentiate them from simple database systems, so there are additional considerations for integrating with HDInsight. You can also integrate at different levels, depending on the way you intend to use the data obtained from HDInsight.

Figure 1-5 shows a sample HDInsight deployment as a data collection and analytics tool.

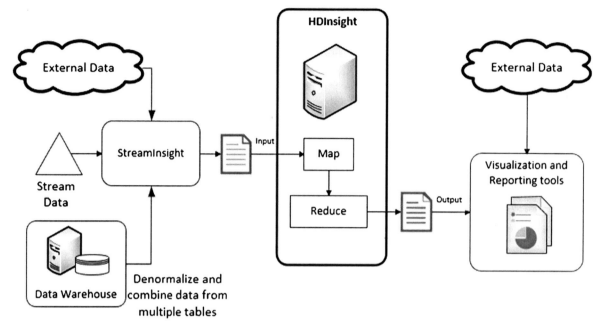

Figure 1-5. *Data collection and analytics*

Enterprise BI is a topic in itself, and there are several factors that require special consideration when integrating a Big Data solution such as HDInsight with an enterprise BI system. You should carefully evaluate the feasibility of integrating HDInsight and the benefits you can get out of it. The ability to combine multiple data sources in a personal data model enables you to have a more flexible approach to data exploration that goes beyond the constraints of a formally managed corporate data warehouse. Users can augment reports and analyses of data from the corporate BI solution with additional data from HDInsight to create a mash-up solution that brings data from both sources into a single, consolidated report.

Figure 1-6 illustrates HDInsight deployment as a powerful BI and reporting tool to generate business intelligence for better decision making.

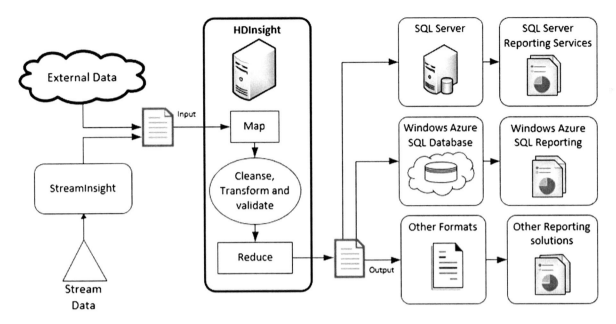

Figure 1-6. *Enterprise BI solution*

Data sources for such models are typically external data that can be matched on a key to existing data in your data store so that it can be used to augment the results of analysis and reporting processes. Following are some examples:

- Social data, log files, sensors, and applications that generate data files

- Datasets obtained from Windows Data Market and other commercial data providers

- Streaming data filtered or processed through SQL Server StreamInsight

---

■ **Note**   Microsoft StreamInsight is a Complex Event Processing (CEP) engine. The engine uses custom-generated events as its source of data and processes them in real time, based on custom query logic (*standing queries* and *events*). The events are defined by a developer/user and can be simple or quite complex, depending on the needs of the business.

You can use the following techniques to integrate output from HDInsight with enterprise BI data at the report level. These techniques are revisited in detail throughout the rest of this book.

- Download the output files generated by HDInsight and open them in Excel, or import them into a database for reporting.

- Create Hive tables in HDInsight, and consume them directly from Excel (including using Power Pivot) or from SQL Server Reporting Services (SSRS) by using the Simba ODBC driver for Hive.

- Use Sqoop to transfer the results from HDInsight into a relational database for reporting. For example, copy the output generated by HDInsight to a Windows Azure SQL Database table and use Windows Azure SQL Reporting Services to create a report from the data.

- Use SQL Server Integration Services (SSIS) to transfer and, if required, transform HDInsight results to a database or file location for reporting. If the results are exposed as Hive tables, you can use an ODBC data source in an SSIS data flow to consume them. Alternatively, you can create an SSIS control flow that downloads the output files generated by HDInsight and uses them as a source for a data flow.

Summary

In this chapter, you saw the different aspects and trends regarding data processing and analytics. Microsoft HDInsight is a collaborative effort with the Apache open source community toward making Apache Hadoop an enterprise-class computing framework that will operate seamlessly, regardless of platform and operating system. Porting the Hadoop ecosystem to Windows, and combining it with the powerful SQL Server Business Intelligence suite of products, opens up different dimensions in data analytics. However, it's incorrect to assume that HDInsight will replace existing database technologies. Instead, it likely will be a perfect complement to those technologies in scenarios that existing RDBMS solutions fail to address.

CHAPTER 2

■ ■ ■

Understanding Windows Azure HDInsight Service

Implementing a Big Data solution is cumbersome and involves significant deployment cost and effort at the beginning to set up the entire ecosystem. It can be a tricky decision for any company to invest such a huge amount of money and resources, especially if that company is merely trying to evaluate a Big Data solution, or if they are unsure of the value that a Big Data solution may bring to the business.

Microsoft offers the Windows Azure HDInsight service as part of an Infrastructure as a Service (IaaS) cloud offering. This arrangement relieves businesses from setting up and maintaining the Big Data infrastructure on their own, so they can focus more on business-specific solutions that execute on the Microsoft cloud data centers. This chapter will provide insight into the various Microsoft cloud offerings and the Windows Azure HDInsight service.

Microsoft's Cloud-Computing Platform

Windows Azure is an enterprise-class, cloud-computing platform that supports both Platform as a Service (PaaS) to eliminate complexity and IaaS for flexibility. IaaS is essentially about getting virtual machines that you must then configure and manage just as you would any hardware that you owned yourself. PaaS essentially gives you preconfigured machines, and really not even machines, but a preconfigured platform having Windows Azure and all the related elements in place and ready for you to use. Thus, PaaS is less work to configure, and you can get started faster and more easily. Use PaaS where you can, and IaaS where you need to.

With Windows Azure, you can use PaaS and IaaS together and independently—you can't do that with other vendors. Windows Azure integrates with what you have, including Windows Server, System Center, Linux, and others. It supports heterogeneous languages, including .NET, Java, Node.js, Python, and data services for No SQL, SQL, and Hadoop. So, if you need to tap into the power of Big Data, simply pair Azure web sites with HDInsight to mine any size data and compelling business analytics to make adjustments to get the best possible business results.

A Windows Azure subscription grants you access to Windows Azure services and to the Windows Azure Management Portal (https://manage.windowsazure.com). The terms of the Windows Azure account, which is acquired through the Windows Azure Account Portal, determine the scope of activities you can perform in the Management Portal and describe limits on available storage, network, and compute resources. A Windows Azure subscription has two aspects:

- The Windows Azure storage account, through which resource usage is reported and services are billed. Each account is identified by a Windows Live ID or corporate e-mail account and associated with at least one subscription. The account owner monitors usage and manages billings through the Windows Azure Account Center.

- The subscription itself, which controls the access and use of Windows Azure subscribed services by the subscription holder from the Management Portal.

13

Figure 2-1 shows you the Windows Azure Management Portal which is your dashboard to manage all your cloud services in one place

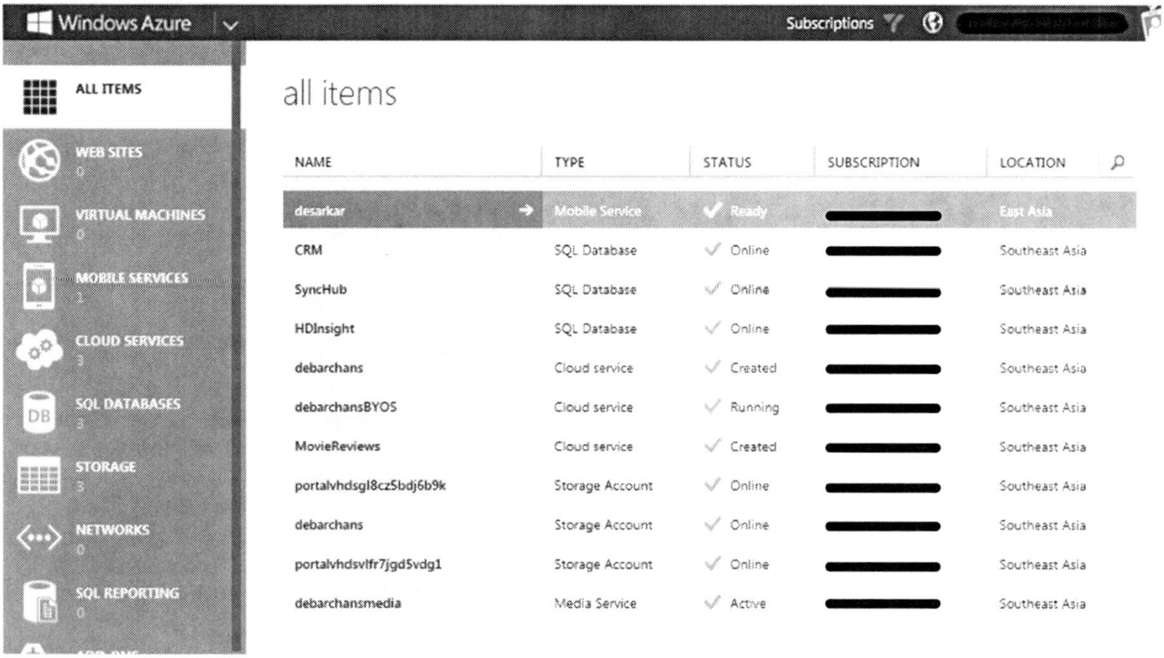

Figure 2-1. *The Windows Azure Management Portal*

The account and the subscription can be managed by the same individual or by different individuals or groups. In a corporate enrollment, an account owner might create multiple subscriptions to give members of the technical staff access to services. Because resource usage within an account billing is reported for each subscription, an organization can use subscriptions to track expenses for projects, departments, regional offices, and so forth.

A detailed discussion of Windows Azure is outside the scope of this book. If you are interested, you should visit the Microsoft official site for Windows Azure:

```
http://www.windowsazure.com/en-us/
```

Windows Azure HDInsight Service

The Windows Azure HDInsight service provides everything you need to quickly deploy, manage, and use Hadoop clusters running on Windows Azure. If you have a Windows Azure subscription, you can deploy your HDInsight clusters using the Azure management portal. Creating your cluster is nothing but provisioning a set of virtual machines in Microsoft Cloud with the Apache Hadoop and its supporting projects bundled in it.

The HDInsight service gives you the ability to gain the full value of Big Data with a modern, cloud-based data platform that manages data of any type, whether structured or unstructured, and of any size. With the HDInsight service, you can seamlessly store and process data of all types through Microsoft's modern data platform that provides simplicity, ease of management, and an open enterprise-ready Hadoop service, all running in the cloud. You can analyze your Hadoop data directly in Excel, using new self-service business intelligence (BI) capabilities like Data Explorer and Power View.

HDInsight Versions

You can choose your HDInsight cluster version while provisioning it using the Azure management dashboard. Currently, there are two versions that are available, but there will be more as updated versions of Hadoop projects are released and Hortonworks ports them to Windows through the Hortonworks Data Platform (HDP).

Cluster Version 2.1

The default cluster version used by Windows Azure HDInsight Service is 2.1. It is based on Hortonworks Data Platform version 1.3.0. It provides Hadoop services with the component versions summarized in Table 2-1.

Table 2-1. - *Hadoop components in HDInsight 2.1*

Component	Version
Apache Hadoop	1.2.0
Apache Hive	0.11.0
Apache Pig	0.11
Apache Sqoop	1.4.3
Apache Oozie	3.2.2
Apache HCatalog	Merged with Hive
Apache Templeton	Merged with Hive
Ambari	API v1.0

Cluster Version 1.6

Windows Azure HDInsight Service 1.6 is another cluster version that is available. It is based on Hortonworks Data Platform version 1.1.0. It provides Hadoop services with the component versions summarized in Table 2-2.

Table 2-2. - *Hadoop components in HDInsight 1.6*

Component	Version
Apache Hadoop	1.0.3
Apache Hive	0.9.0
Apache Pig	0.9.3
Apache Sqoop	1.4.2
Apache Oozie	3.2.0
Apache HCatalog	0.4.1
Apache Templeton	0.1.4
SQL Server JDBC Driver	3.0

■ **Note** Both versions of the cluster ship with stable components of HDP and the underlying Hadoop eco-system. However, I recommend the latest version, which is 2.1 as of this writing. The latest version will have the latest enhancements and updates from the open source community. It will also have fixes to bugs that were reported against previous versions. For those reasons, my preference is to run on the latest available version unless there is some specific reason to do otherwise by running some older version.

The component versions associated with HDInsight cluster versions may change in future updates to HDInsight. One way to determine the available components and their versions is to login to a cluster using Remote Desktop, go directly to the cluster's name node, and then examine the contents of the *C:\apps\dist* directory.

Storage Location Options

When you create a Hadoop cluster on Azure, you should understand the different storage mechanisms. Windows Azure has three types of storage available: blob, table, and queue:

- **Blob storage:** Binary Large Objects (blob) should be familiar to most developers. Blob storage is used to store things like images, documents, or videos—something larger than a first name or address. Blob storage is organized by containers that can have two types of blobs: Block and Page. The type of blob needed depends on its usage and size. Block blobs are limited to 200 GBs, while Page blobs can go up to 1 TB. Blob storage can be accessed via REST APIs with a URL such as http://debarchans.blob.core.windows.net/MyBLOBStore.

- **Table storage:** Azure tables should not be confused with tables from an RDBMS like SQL Server. They are composed of a collection of entities and properties, with properties further containing collections of name, type, and value. One thing I particularly don't like as a developer is that Azure tables can't be accessed using ADO.NET methods. As with all other Azure storage methods, access is provided through REST APIs, which you can access at the following site: http://debarchans.table.core.winodws.net/MyTableStore.

- **Queue storage:** Queues are used to transport messages between applications. Azure queues are conceptually the same as Microsoft Messaging Queue (MSMQ), except that they are for the cloud. Again, REST API access is available. For example, this could be an URL like: http://debarchans.queue.core.windows.net/MyQueueStore.

■ **Note** HDInsight supports only Azure blob storage.

Azure storage accounts

The HDInsight provision process requires a Windows Azure Storage account to be used as the default file system. The storage locations are referred to as Windows Azure Storage Blob (WASB), and the acronym *WASB:* is used to access them. WASB is actually a thin wrapper on the underlying Windows Azure Blob Storage (WABS) infrastructure, which exposes blob storage as HDFS in HDInsight and is a notable change in Microsoft's implementation of Hadoop on Windows Azure. (Learn more about WASB in the upcoming section *Understanding the Windows Azure Storage Blob*). For instructions on creating a storage account, see the following URL:

http://www.windowsazure.com/en-us/manage/services/storage/how-to-create-a-storage-account/

The HDInsight service provides access to the distributed file system that is locally attached to the compute nodes. This file system can be accessed using the fully qualified URI—for example:

```
hdfs://<namenode>/<path>
```

The syntax to access WASB is

```
WASB://[<container>@]<accountname>.blob.core.windows.net/<path>
```

Hadoop supports the notion of a default file system. The default file system implies a default scheme and authority; it can also be used to resolve relative paths. During the HDInsight provision process, you must specify blob storage and a container used as the default file system to maintain compatibility with core Hadoop's concept of default file system. This action adds an entry to the configuration file *C:\apps\dist\hadoop-1.1.0-SNAPSHOT\conf\core-site.xml* for the blob store container.

■ **Caution**   Once a storage account is chosen, it cannot be changed. If the storage account is removed, the cluster will no longer be available for use.

Accessing containers

In addition to accessing the blob storage container designated as the default file system, you can also access containers that reside in the same Windows Azure storage account or different Windows Azure storage accounts by modifying *C:\apps\dist\hadoop-1.1.0-SNAPSHOT\conf\core-site.xml* and adding additional entries for the storage accounts. For example, you can add entries for the following:

- **Container in the same storage account:** Because the account name and key are stored in the core-site.xml during provisioning, you have full access to the files in the container.

- **Container in a different storage account with the public container or the public blob access level:** You have read-only permission to the files in the container.

- **Container in a different storage account with the private access levels:** You must add a new entry for each storage account to the *C:\apps\dist\hadoop-1.1.0-SNAPSHOT\conf\core-site.xml* file to be able to access the files in the container from HDInsight, as shown in Listing 2-1.

Listing 2-1. Accessing a Blob Container from a Different Storage Account

```
<property>
<name>fs.azure.account.key.<YourStorageAccountName>.blob.core.microsoft.com</name>
<value><YourStorageAccountkeyValue></value>
</property>
```

■ **Caution**   Accessing a container from another storage account might take you outside of your subscription's data center. You might incur additional charges for data flowing across the data-center boundaries.

Understanding the Windows Azure Storage Blob

HDInsight introduces the unique Windows Azure Storage Blob (WASB) as the storage media for Hadoop on the cloud. As opposed to the native HDFS, the Windows Azure HDInsight service uses WASB as its default storage for the Hadoop clusters. WASB uses Azure blob storage underneath to persist the data. Of course, you can choose to override the defaults and set it back to HDFS, but there are some advantages to choosing WASB over HDFS:

- WASB storage incorporates all the HDFS features, like fault tolerance, geo replication, and partitioning.

- If you use WASB, you disconnect the data and compute nodes. That is not possible with Hadoop and HDFS, where each node is both a data node and a compute node. This means that if you are not running large jobs, you can reduce the cluster's size and just keep the storage—and probably at a reduced cost.

- You can spin up your Hadoop cluster only when needed, and you can use it as a "transient compute cluster" instead of as permanent storage. It is not always the case that you want to run idle compute clusters to store data. In most cases, it is more advantageous to create the compute resources on-demand, process data, and then de-allocate them without losing your data. You cannot do that in HDFS, but it is already done for you if you use WASB.

- You can spin up multiple Hadoop clusters that crunch the same set of data stored in a common blob location. In doing so, you essentially leverage Azure blob storage as a shared data store.

- Storage costs have been benchmarked to approximately five times lower for WASB than for HDFS.

- HDInsight has added significant enhancements to improve read/write performance when running Map/Reduce jobs on the data from the Azure blob store.

- You can process data directly, without importing to HDFS first. Many people already on a cloud infrastructure have existing pipelines, and those pipelines can push data directly to WASB.

- Azure blob storage is a useful place to store data across diverse services. In a typical case, HDInsight is a piece of a larger solution in Windows Azure. Azure blob storage can be the common link for unstructured blob data in such an environment.

■ **Note** Most HDFS commands—such as `ls`, `copyFromLocal`, and `mkdir`—will still work as expected. Only the commands that are specific to the native HDFS implementation (which is referred to as *DFS*), such as `fschk` and `dfsadmin`, will show different behavior on WASB.

Figure 2-2 shows the architecture of an HDInsight service using WASB.

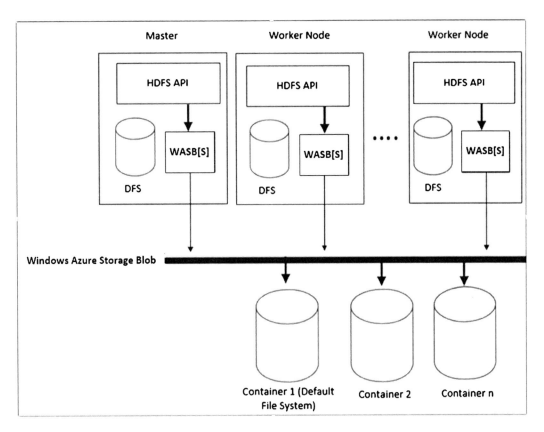

Figure 2-2. *HDInsight with Azure blob storage*

As illustrated in Figure 2-2, the master node as well as the worker nodes in an HDInsight cluster default to WASB storage, but they also have the option to fall back to traditional DFS. In the case of default WASB, the nodes, in turn, use the underlying containers in the Windows Azure blob storage.

Uploading Data to Windows Azure Storage Blob

Windows Azure HDInsight clusters are typically deployed to execute MapReduce jobs, and are dropped once these jobs have completed. Retaining large volumes data in HDFS after computations are done is not at all cost effective. Windows Azure Blob Storage is a highly available, scalable, high capacity, low cost, and shareable storage option for data that is to be processed using HDInsight. Storing data in WASB enables your HDInsight clusters to be independent of the underlying storage used for computation, and you can safely release those clusters without losing data.

The first step toward deploying an HDInsight solution on Azure is to decide on a way to upload data to WASB efficiently. We are talking BigData here. Typically, the data that needs to be uploaded for processing will be in the terabytes and petabytes. This section highlights some off-the-shelf tools from third-parties that can help in uploading such large volumes to WASB storage. Some of the tools are free, and some you need to purchase.

> **Azure Storage Explorer:** A free tool that is available from codeplex.com. It provides a nice Graphical User Interface from which to manage your Azure Blob containers. It supports all three types of Azure storage: blobs, tables, and queues. This tool can be downloaded from:
>
> http://azurestorageexplorer.codeplex.com/

Cloud Storage Studio 2: This is a paid tool giving you complete control of your Windows Azure blobs, tables, and queues. You can get a 30-day trial version of the tool from here:

```
http://www.cerebrata.com/products/cloud-storage-studio/introduction
```

CloudXplorer: This is also a paid tool available for Azure storage management. Although the release versions of this tool need to be purchased, there is still an older version available as freeware. That older version can be downloaded from the following URL:

```
http://clumsyleaf.com/products/cloudxplorer
```

Windows Azure Explorer: This is another Azure storage management utility which offers both a freeware and a paid version. A 30-day trial of the paid version is available. It is a good idea to evaluate either the freeware version or the 30-day trial before making a purchase decision. You can grab this tool from the following page:

```
http://www.cloudberrylab.com/free-microsoft-azure-explorer.aspx
```

Apart from these utilities, there are a few programmatic interfaces that enable you to develop your own application to manage your storage blobs. Those utilites are:

- AzCopy
- Windows Azure PowerShell
- Windows Azure Storage Client Library for .NET
- Hadoop command line

To get a complete understanding on how you can implement these programmatic interfaces and build your own data upload solution, check the link below:

```
http://www.windowsazure.com/en-us/manage/services/hdinsight/howto-upload-data-to-hdinsight/
```

Windows Azure Flat Network Storage

Traditional Hadoop leverages the locality of data per node through HDFS to reduce data traffic and network bandwidth. On the other hand, HDInsight promotes the use of WASB as the source of data, thus providing a unified and more manageable platform for both storage and computation, which makes sense. But an obvious question that comes up regarding this architecture is this: Will this setup have a bigger network bandwidth cost? The apparent answer seems to be "Yes," because the data in WASB is no longer local to the compute nodes. However, the reality is a little different.

Overall, when using WASB instead of HDFS you should not encounter performance penalties. HDInsight ensures that the Hadoop cluster and storage account are co-located in the same flat data center network segment. This is the next-generation data-center networking architecture also referred to as the "Quantum 10" (Q10) network architecture. Q10 architecture flattens the data-center networking topology and provides full bisection bandwidth between compute and storage. Q10 provides a fully nonblocking, 10-Gbps-based, fully meshed network, providing an aggregate backplane in excess of 50 Tbps of bandwidth for each Windows Azure data center. Another major improvement in reliability and throughput is moving from a hardware load balancer to a software load balancer. This entire architecture is based on a research paper by Microsoft, and the details can be found here:

```
http://research.microsoft.com/pubs/80693/vl2-sigcomm09-final.pdf
```

In the year 2012, Microsoft deployed this flat network for Windows Azure across all of the data centers to create Flat Network Storage (FNS). The result is very high bandwidth network connectivity for storage clients. This new network design enables MapReduce scenarios that can require significant bandwidth between compute and storage. Microsoft plans to continue to invest in improving bandwidth between compute and storage, as well as increase the scalability targets of storage accounts and partitions as time progresses. Figure 2-3 shows a conceptual view of Azure FNS interfacing between blob storage and the HDInsight compute nodes.

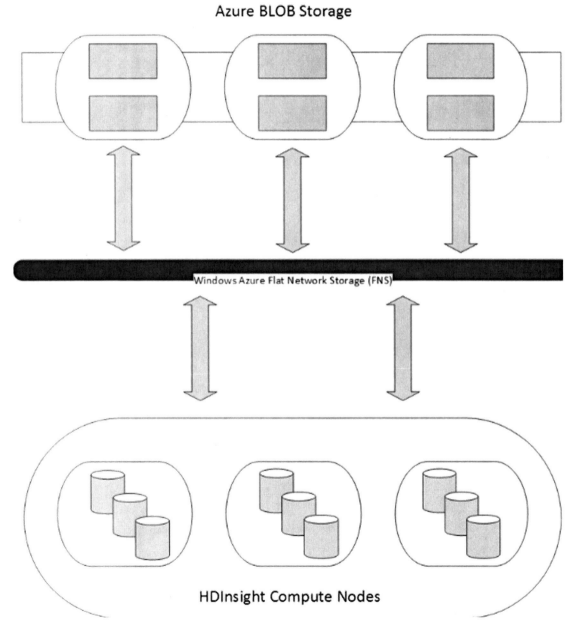

Figure 2-3. *Azure Flat Network Storage*

Summary

In this chapter, you read about the Windows Azure HDInsight service. You had a look into subscribing to the HDInsight service, which defaults to the Windows Windows Azure Storage Blob (WASB) as the data repository rather than to HDFS as in traditional Hadoop. This chapter covered the benefits of using WASB as the storage media in the cloud, and it mentioned some available tools for uploading data to WASB. Also discussed was the brand new Azure Flat Network Storage (FNS) designed specifically for improved network bandwidth and throughput.

## CHAPTER 3

Provisioning Your HDInsight Service Cluster

The HDInsight Service brings you the simplicity of deploying and managing your Hadoop clusters in the cloud, and it enables you to do that in a matter of just a few minutes. Enterprises can now free themselves of the considerable cost and effort of configuring, deploying, and managing Hadoop clusters for their data-mining needs. As a part of its Infrastructure as a Service (IaaS) offerings, HDInsight also provides a cost-efficient approach to managing and storing data. The HDInsight Service uses Windows Azure blob storage as the default file system.

Note   An Azure storage account is required to provision a cluster. The storage account you associate with your cluster is where you will store the data that you will analyze in HDInsight.

Creating the Storage Account

You can have multiple storage accounts under your Azure subscription. You can choose any of the existing storage accounts you already have where you want to persist your HDInsight clusters' data, but it is always a good practice to have dedicated storage accounts for each of your Azure services. You can even choose to have your storage accounts in different data centers distributed geographically to reduce the impact on the rest of the services in the unlikely event that a data center goes down.

To create a storage account, log on to your Windows Azure Management Portal (https://manage.windowsazure.com) and navigate to the storage section as shown in Figure 3-1.

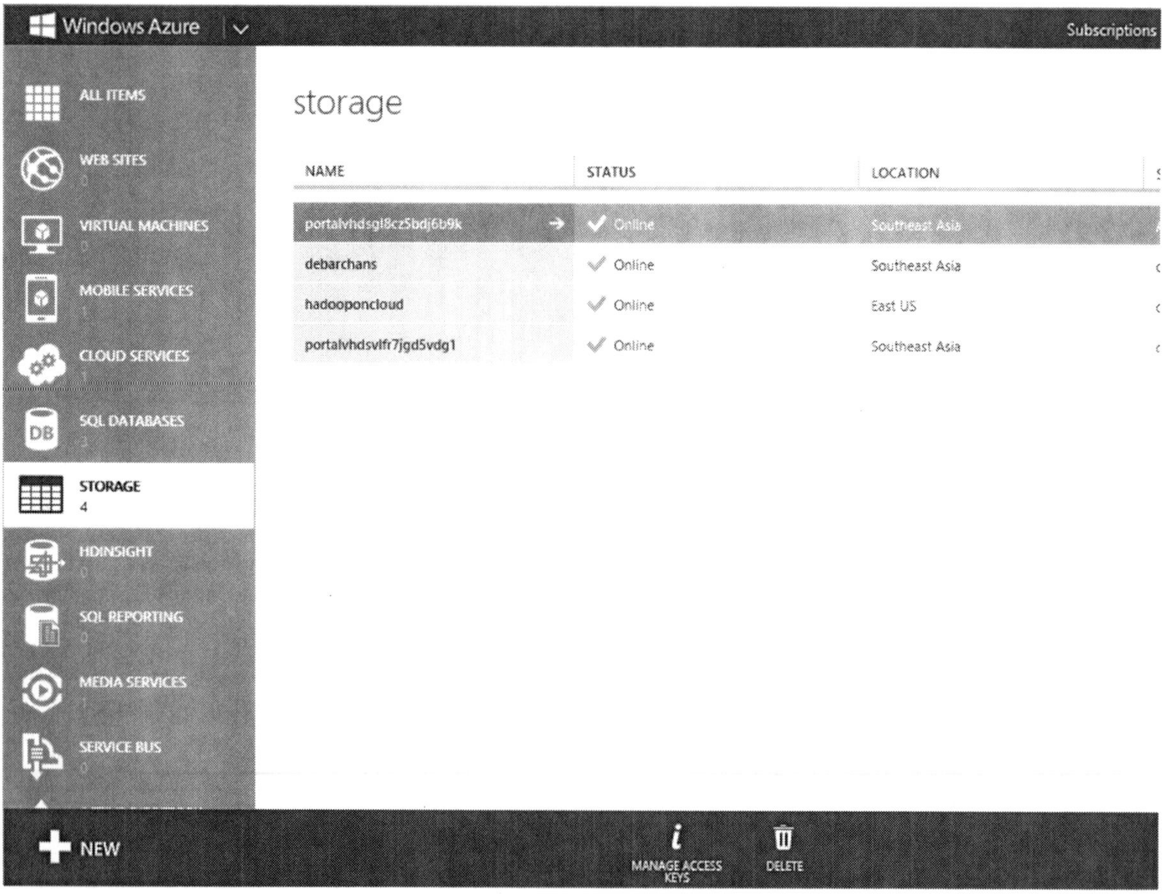

Figure 3-1. *Windows Azure Management Portal*

■ **Note** You might need to provide your Azure subscription credentials the first time you try to access the Management Portal.

Click on the NEW button in the lower left corner to bring up the NEW ➤ DATA SERVICES ➤ STORAGE window as shown in Figure 3-2.

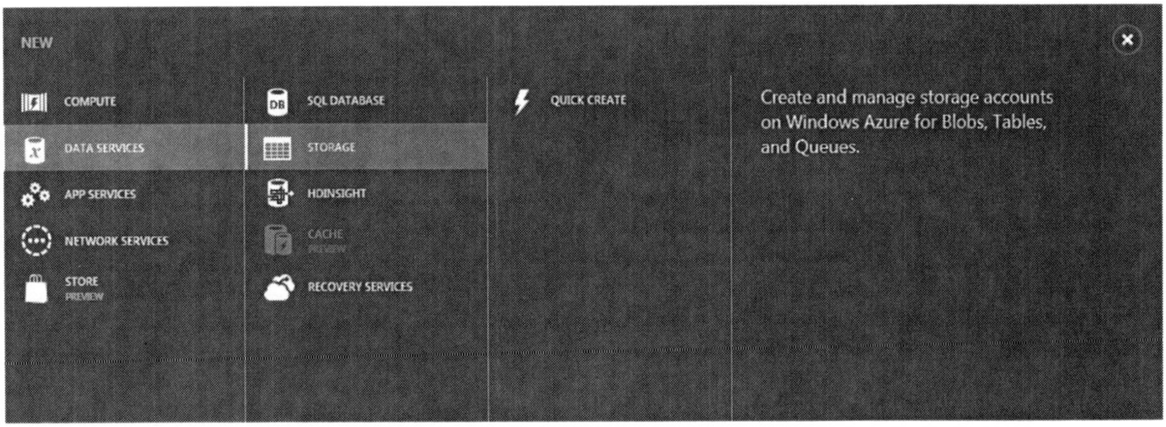

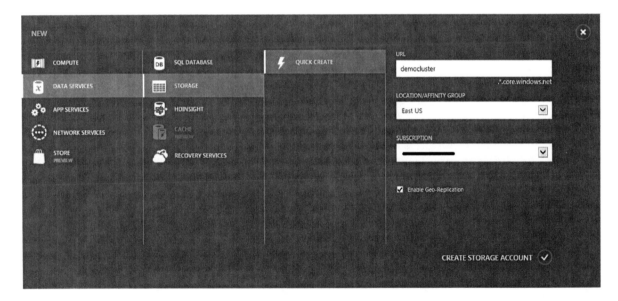

Figure 3-2. *New storage account*

Click on QUICK CREATE. Provide the storage account name, and select the location of the data-center region. If you have multiple subscriptions, you can also choose to select the one that gets billed according to your usage of the storage account. After providing all these details, your screen should look like Figure 3-3.

Figure 3-3. *Storage account details*

If you wish, Windows Azure can geo-replicate your Windows Azure Blob and Table data, at no additional cost, between two locations hundreds of miles apart within the same region (for example, between North and South US, between North and West Europe, and between East and Southeast Asia). Geo-replication is provided for additional data durability in case of a major data-center disaster. Select the Enable Geo-Replication check box if you want that functionality enabled. Then click on CREATE STORAGE ACCOUNT to complete the process of adding a storage account. Within a minute or two, you should see the storage account created and ready for use in the portal as shown in Figure 3-4.

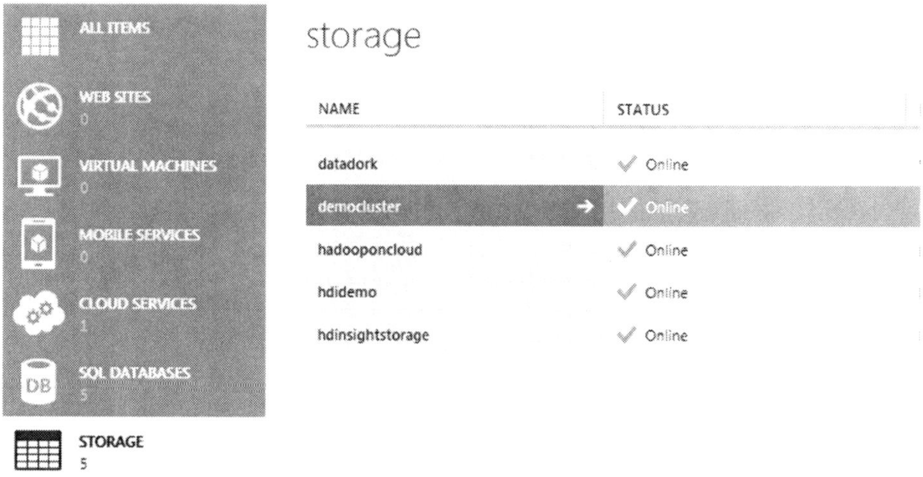

NAME	STATUS
datadork	✓ Online
democluster →	✓ Online
hadooponcloud	✓ Online
hdidemo	✓ Online
hdinsightstorage	✓ Online

Figure 3-4. *The democluster storage account*

■ **Note** Enabling geo-replication later for a storage account that has data in it might have a pricing impact on the subscription.

Creating a SQL Azure Database

When you actually provision your HDInsight cluster, you also get the option of customizing your Hive and Oozie data stores. In contrast to the traditional Apache Hadoop, HDInsight gives you the option of selecting a SQL Azure database for storing the metadata for Hive and Oozie. This section quickly explains how to create a SQL database on Azure, which you would later use as storage for Hive and Oozie.

Create a new SQL Azure database from your Azure Management Portal. Click on New ➤ Data Services ➤ SQL Database. Figure 3-5 shows the use of the QUICK CREATE option to create the database.

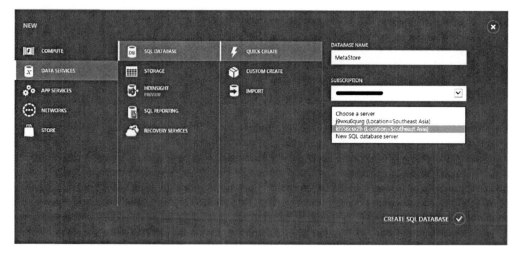

Figure 3-5. *Creating a SQL Azure database*

The choices in Figure 3-5 will create a database on your Azure data center with the name MetaStore. It will be 1 GB in size, and it should be listed in your Azure portal as shown in Figure 3-6.

MetaStore	SQL Database	✓ Online
democluster	HDInsight Cluster	✓ Running

Figure 3-6. *The MetaStore SQL Azure database*

You can further customize your database creation by specifying the database size, collation, and more using the CUSTOM CREATE option instead of the QUICK CREATE option. (You can see CUSTOM CREATE just under QUICK CREATE in Figure 3-5). You can even import an existing database backup and restore it as a new database using the IMPORT option in the wizard.

However you choose to create it, you now have a database in SQL Azure. You will later use this database as metadata storage for Hive and Oozie when you provision your HDInsight cluster.

Deploying Your HDInsight Cluster

Now that you have your dedicated storage account ready, select the HDINSIGHT option in the portal and click on CREATE AN HDINSIGHT CLUSTER as shown in Figure 3-7.

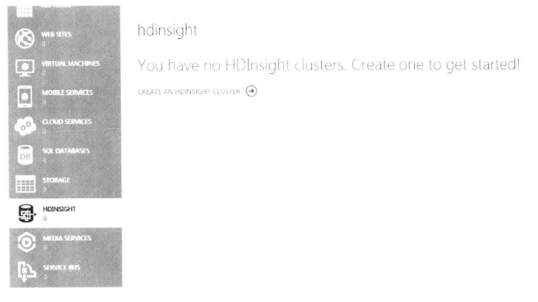

Figure 3-7. *Create new HDInsight cluster*

Click on QUICK CREATE to bring up the cluster configuration screen. Provide the name of your cluster, choose the number of data nodes, and select the storage account democluster that was created earlier as the default storage account for your cluster, as shown in Figure 3-8. You must also provide a cluster user password. The password must be at least 10 characters long and must contain an uppercase letter, a lowercase letter, a number, and a special character.

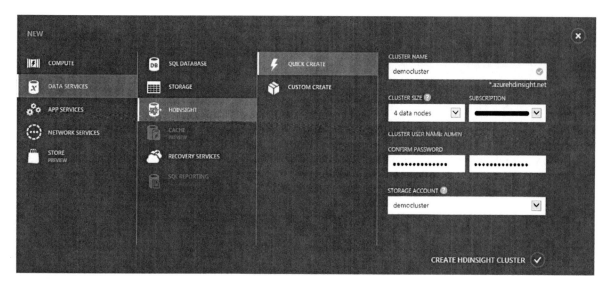

Figure 3-8. *HDInsight cluster details*

■ **Note**   You can select the number of data nodes between options to be 4, 8, 16, or 32. Any number of data nodes can be specified when using the CUSTOM CREATE option discussed in the next section. Pricing details on the billing rates for various cluster sizes are available. Click on the *?* symbol just above the drop-down box, and follow the link on the popup.

Customizing Your Cluster Creation

You can also choose CUSTOM CREATE to customize your cluster creation further. Clicking on CUSTOM CREATE launches a three-step wizard. The first step requires you to provide the cluster name and specify the number of nodes, as shown in Figure 3-9. You can specify your data-center region and any number of nodes here, unlike the fixed set of options available with QUICK CREATE.

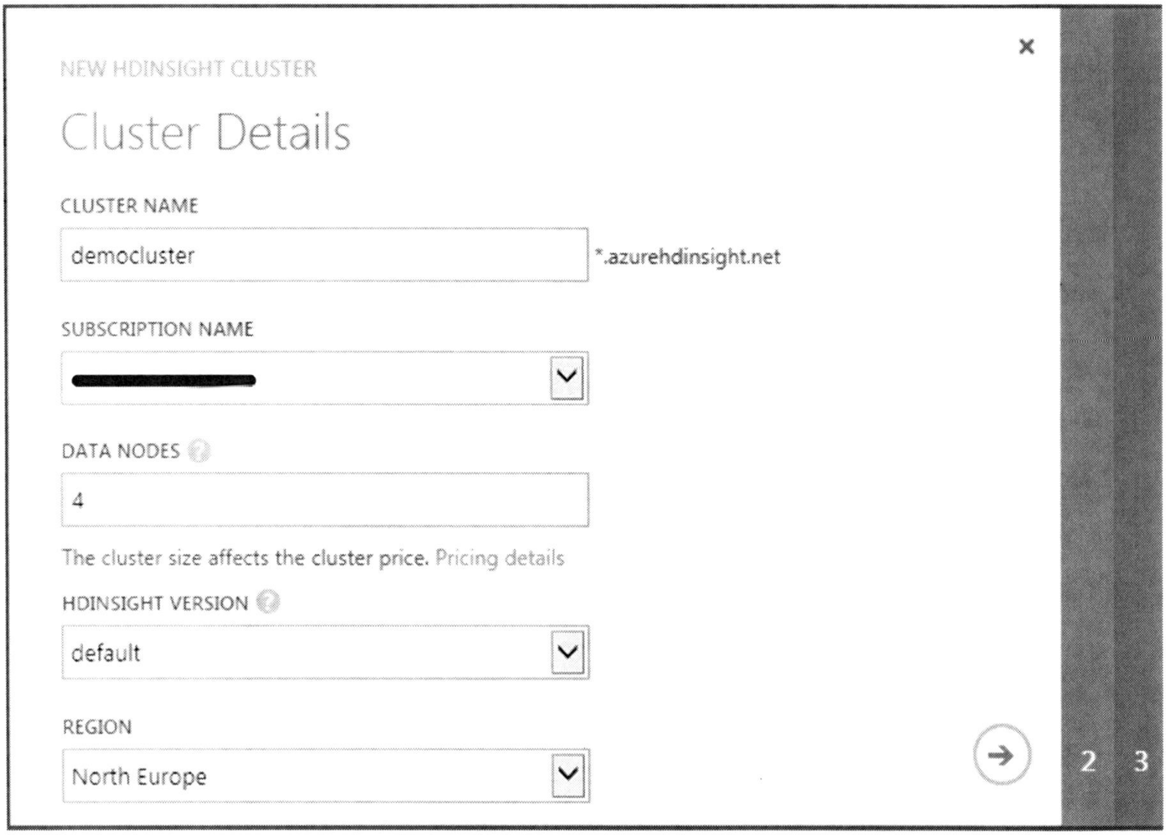

Figure 3-9. *Customizing the cluster creation*

Configuring the Cluster User and Hive/Oozie Storage

Click on the Next arrow in the bottom right corner of the wizard to bring up the Configure Cluster User screen. Provide the cluster credentials you would like to be set for accessing the HDInsight cluster. Here, you can specify the Hive/Oozie metastore to be the SQL Azure database you just created, as shown in Figure 3-10.

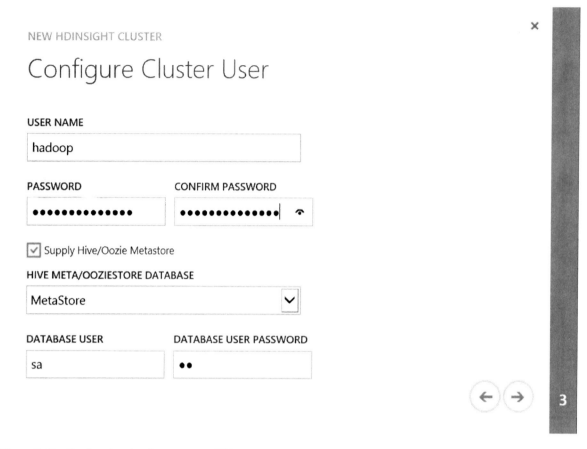

NEW HDINSIGHT CLUSTER

Configure Cluster User

USER NAME

hadoop

PASSWORD   **CONFIRM PASSWORD**

●●●●●●●●●●●●●●   ●●●●●●●●●●●●●●|

☑ Supply Hive/Oozie Metastore

HIVE META/OOZIESTORE DATABASE

MetaStore

DATABASE USER   **DATABASE USER PASSWORD**

sa   ●●

Figure 3-10. *Configuring the cluster user and Hive metastore*

■ **Note**   If you choose QUICK CREATE to create your cluster, the default user name is Admin. This can be changed only by using the CUSTOM CREATE wizard.

By default Hive/Oozie uses an open source RDBMS system for its storage called Derby. It can be embedded in a Java program (like Hive), and it supports online transaction processing. If you wish to continue with Derby for your Hive and Oozie storage, you can choose to leave the box deselected.

Choosing Your Storage Account

The next step of the wizard is to select the storage account for the cluster. You can use the already created *democluster* account to associate with the cluster. You also get an option here to create a dedicated storage account on the fly or even to use a different storage account from a different subscription altogether. This step also gives you the option of creating a default container in the storage account on the fly, as shown in Figure 3-11. Be careful, though, because once as storage account for the cluster is chosen, it cannot be changed. If the storage account is removed, the cluster will no longer be available for use.

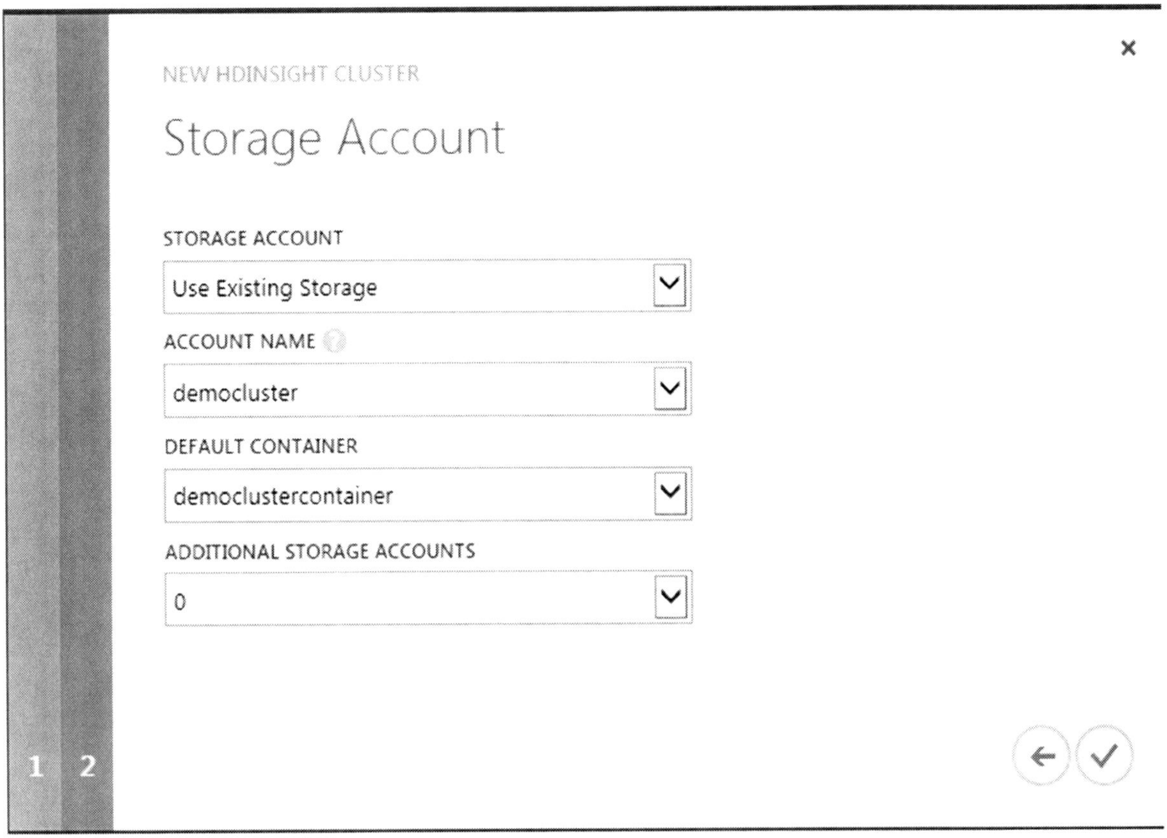

Figure 3-11. Specifying the HDInsight cluster storage account

■ **Note** The name of the default container is the same name as that of the HDInsight cluster. In this case, I have pre-created my container in the storage account which is *democlustercontainer*.

The CUSTOM CREATE wizard also gives you the option to specify multiple storage accounts for your cluster. The wizard provides you additional storage account configuration screens in case you provide a value for the ADDITIONAL STORAGE ACCOUNTS drop-down box as shown in Figure 3-11. For example, if you wish to associate two more storage accounts with your cluster, you can select the value 2, and there will be two more additional screens in the wizard as shown in Figure 3-12.

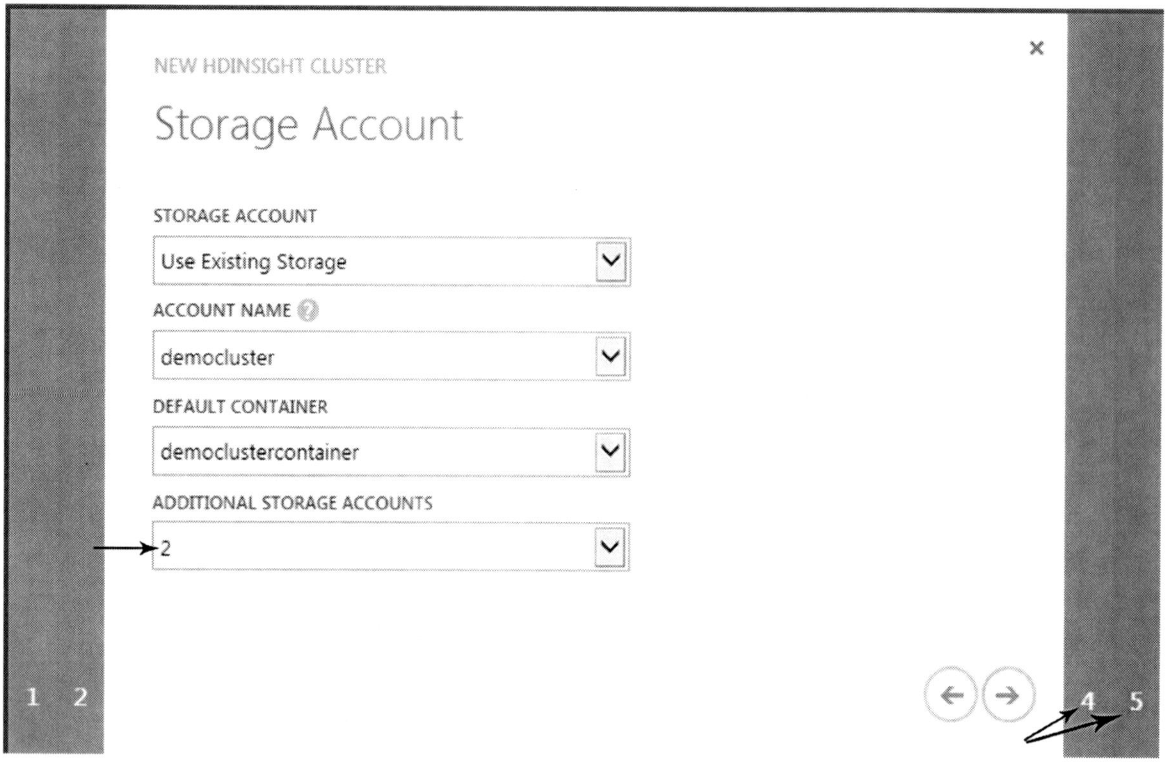

Figure 3-12. Adding more storage accounts

Finishing the Cluster Creation

Click on Finish (the check mark button) to complete the cluster-creation process. It will take up to several minutes to provision the name node and the data nodes, depending on your chosen configuration, and you will see several status messages like one shown in Figure 3-13 throughout the process.

Figure 3-13. Cluster creation in process

Eventually, the cluster will be provisioned. When it is available, its status is listed as Running, as shown in Figure 3-14.

NAME	STATUS	SUBSCRIPTION NAME	LOCATION
democluster	→ ✓ Running	━━━━━━━	East US

Figure 3-14. *An HDInsight cluster that's ready for use*

Monitoring the Cluster

You can click on *democluster*, which you just created, to access your cluster dashboard. The dashboard provides a quick glance of the metadata for the cluster. It also gives you an overview of the entire cluster configuration, its usage, and so on, as shown in Figure 3-15. At this point, your cluster is fresh and clean. We will revisit the dashboard later, after the cluster is somewhat active, and check out the differences.

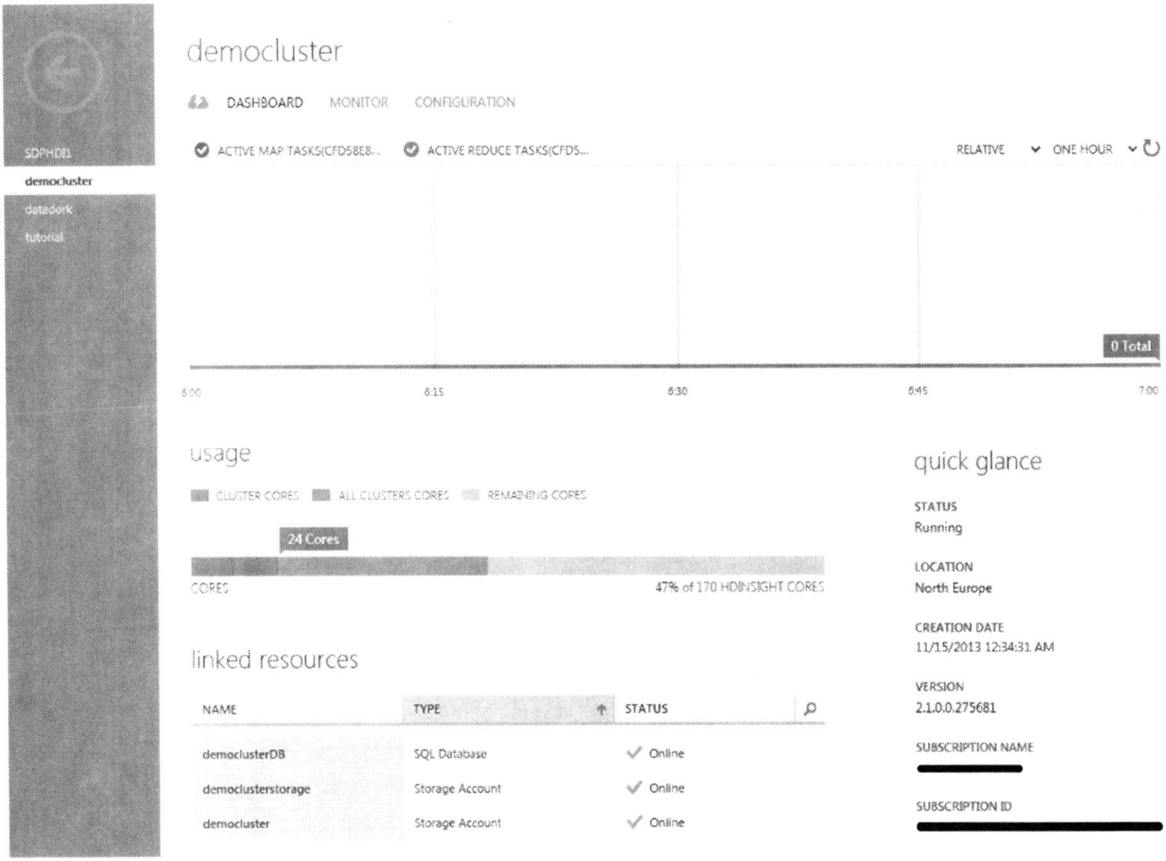

Figure 3-15. *The HDInsight cluster dashboard*

You can also click the MONITOR option to have a closer look at the currently active mappers and reducers, as shown in Figure 3-16. Again, we will come back to this screen later while running a few map-reduce jobs on the cluster.

Figure 3-16. *Monitoring your cluster*

You can also choose to alter the filters and customize the refresh rate for the dashboard, as shown in Figure 3-17.

Figure 3-17. *Setting the dashboard refresh rate*

Configuring the Cluster

If you want to control the Hadoop services running on the name node, you can do that from the Configuration tab as shown in Figure 3-18.

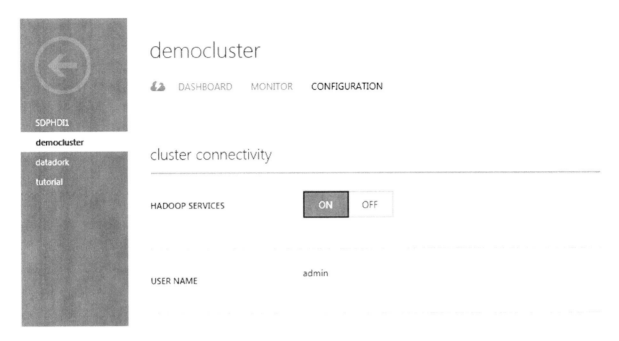

Figure 3-18. Configuring Hadoop services

Hadoop services are turned on by default. You can click the OFF button to stop the services in the name node. You can also enable Remote Desktop access to your name node from the Configuration screen. Do that through the ENABLE REMOTE button at the bottom of this screen, as shown in Figure 3-19.

Figure 3-19. Enable Remote Desktop

Once you click on ENABLE REMOTE, you get an option to configure a remote user. Specify the password and a date when the remote access permission expires. The expiration is for security reasons. It forces you to periodically visit this configuration screen and extend the remote access privilege, so that it doesn't remain past when it is needed. Figure 3-20 shows the remote user configuration screen.

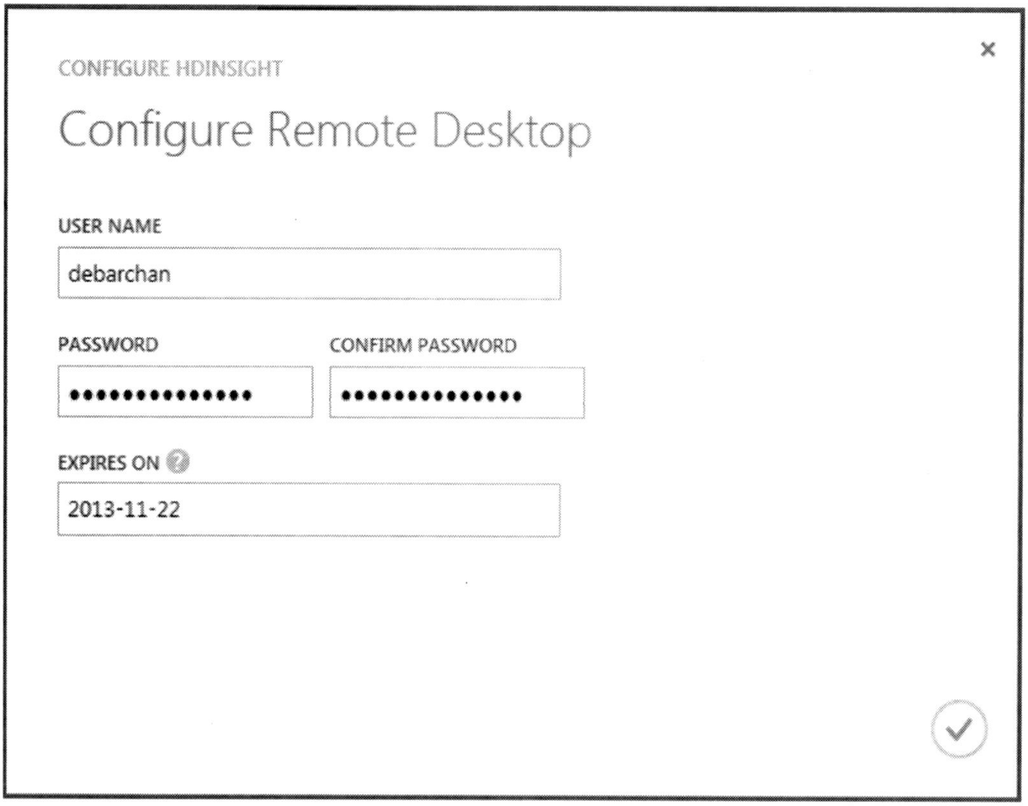

CONFIGURE HDINSIGHT

Configure Remote Desktop

USER NAME

debarchan

PASSWORD

••••••••••••••

CONFIRM PASSWORD

••••••••••••••

EXPIRES ON ⑦

2013-11-22

Figure 3-20. *Configure Remote Desktop*

Once Remote Desktop is configured for the cluster, you should see status messages similar to those in Figure 3-21.

Figure 3-21. *Remote Desktop is enabled*

You can come back to the cluster configuration screen anytime you wish to disable Remote Desktop access. Do that via the DISABLE REMOTE button shown in Figure 3-22.

Figure 3-22. *Disable Remote Desktop*

Once you are done with your cluster, you can choose to delete the cluster by pressing the DELETE button in the configuration screen. Figure 3-22 shows that button too.

Once the cluster deletion process is complete, you will see status messages similar to those in Figure 3-23.

Figure 3-23. *Deleting the cluster*

Summary

This chapter gets you started using the Windows Azure HDInsight Service, which makes Apache Hadoop available as a service in the cloud. You saw how to provision your Hadoop clusters in the cloud using the simple wizards available in the Azure Management Portal. You also saw how to create a dedicated storage account and associate it with the cluster that is used as the default file system by HDInsight.

CHAPTER 4

■ ■ ■

Automating HDInsight Cluster Provisioning

It is almost always a requirement for a business to automate activities that are repetitive and can be predicted well in advance. Through the strategic use of technology and automation, an organization can increase its productivity and efficiency by automating recurring tasks associated with the daily workflow. Apache Hadoop exposes Java interfaces for developers to programmatically manipulate and automate the creation of Hadoop clusters.

Microsoft .NET Framework is part of the automation picture in HDInsight. Existing .NET developers can now leverage their skillset to automate workflows in the Hadoop world. Programmers now have the option to write their MapReduce jobs in C# and VB .NET. Additionally, HDInsight also supports Windows PowerShell to automate cluster operations through scripts. PowerShell is a script-based workflow and is a particular favorite of Windows administrators for scripting their tasks. There is also a command-based interface based on Node.js to automate cluster-management operations. This chapter will discuss the various ways to use the Hadoop .NET Software Development Kit (SDK), Windows PowerShell, and the cross-platform Command-Line Interface (CLI) tools to automate HDInsight service cluster operations.

Using the Hadoop .NET SDK

The Hadoop .NET SDK provides .NET client API libraries that make it easier to work with Hadoop from .NET. Since all of this is open source, the SDK is hosted in the open source site CodePlex and can be downloaded from the following link:

```
http://hadoopsdk.codeplex.com/
```

CodePlex uses NuGet packages to help you easily incorporate components for certain functions. NuGet is a Visual Studio extension that makes it easy to add, remove, and update libraries and tools in Visual Studio projects that use the .NET Framework. When you add a library, NuGet copies files to your solution and automatically adds and updates the required references in your *app.config* or *web.config* file. NuGet also makes sure that it reverts those changes when the library is dereferenced from your project so that nothing is left behind. For more detailed information, visit the NuGet documentation site:

```
http://nuget.codeplex.com/documentation
```

There are NuGet packages for HDInsight that need to be added to your solution. Starting with Visual Studio 2013, the version that I am using to build the samples for this book, NuGet is included in every edition (except Team Foundation Server) by default. If you are developing on a Visual Studio 2010 platform or for some reason you cannot find it in Visual Studio 2013, you can download the extension from the following link.

```
http://docs.nuget.org/docs/start-here/installing-nuget
```

Once you download the extension, you will have a NuGet.Tools.vsix file, which is a Visual Studio Extension. Execute the file and the VSIX installer will install the Visual Studio add-in. Note that you will need to restart Visual Studio if it is already running after the add-in installation. This add-in will enable you to import the NuGet packages for HDInsight in your Visual Studio application.

Adding the NuGet Packages

To use the HDInsight NuGet packages, you need to create a solution first. Since we are going perform the cluster-management operations that we can see from the Azure portal, a console application is good enough to demonstrate the functionalities. Launch Visual Studio 2013, and choose to create a new C# Console Application from the list of available project types, as shown in Figure 4-1.

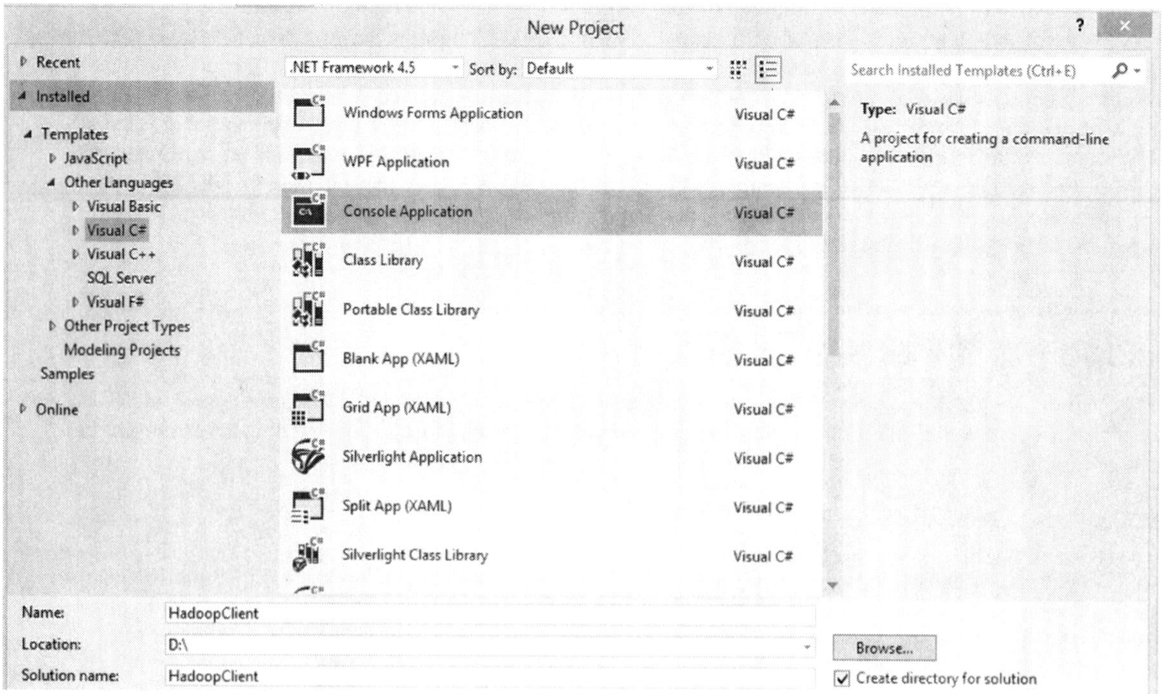

Figure 4-1. *New C# console application*

Once the solution is created, open the NuGet *Package Manager Console* to import the required packages, as shown in Figure 4-2.

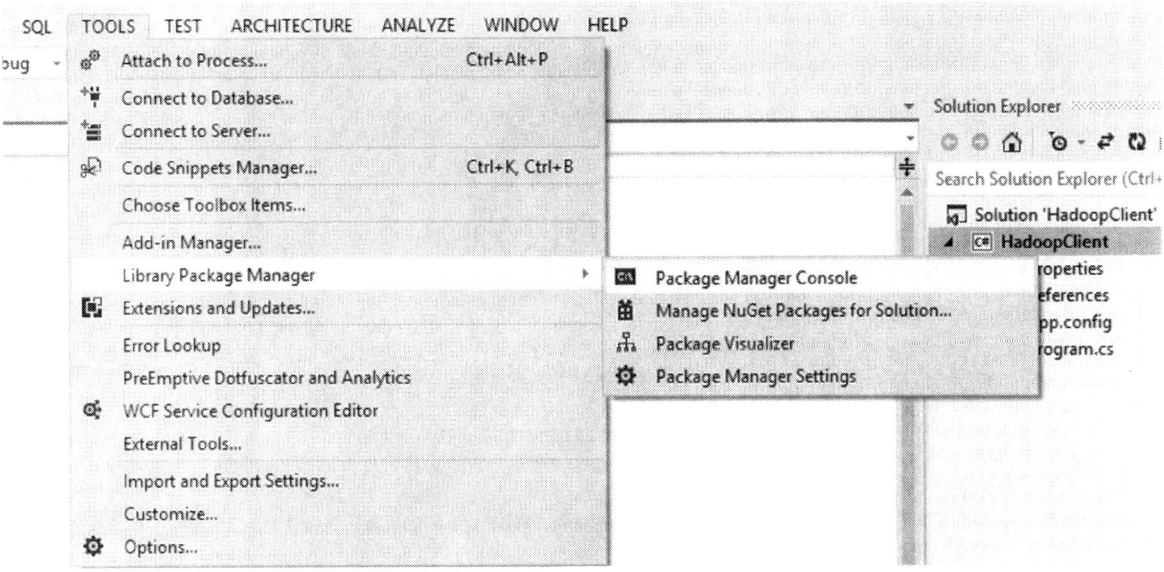

Figure 4-2. *NuGet Package Manager Console*

Table 4-1 summarizes the NuGet packages available currently for HDInsight, with a brief description of each.

Table 4-1. *HDInsight NuGet packages*

Package Name	Function
Microsoft.WindowsAzure.Management.HDInsight	Set of APIs for HDInsight cluster-management operations.
Microsoft.Hadoop.WebClient	Set of APIs to work with the Hadoop file system.
Microsoft.Hadoop.Hive	Set of APIs for Hive operations.
Microsoft.Hadoop.MapReduce	Set of APIs for MapReduce job submission and execution.
Microsoft.Hadoop.Avro	Set of APIs for data serialization, based on the Apache open source project Avro.

In your HadoopClient solution, install the `Microsoft.WindowsAzure.Management.HDInsight` package by running the following command in the Package Manager Console:

```
install-package Microsoft.WindowsAzure.Management.HDInsight
```

Figure 4-3 shows how you would type the command into the Visual Studio Package Manager Console.

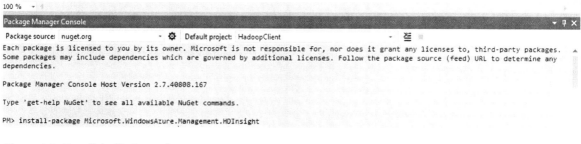

Figure 4-3. *Install the NuGet package*

You should see the following output if the package is imported successfully:

```
Installing 'Microsoft.WindowsAzure.Management.HDInsight 0.9.4951.25594'.
Successfully installed 'Microsoft.WindowsAzure.Management.HDInsight 0.9.4951.25594'.
Adding 'Microsoft.WindowsAzure.Management.HDInsight 0.9.4951.25594' to HadoopClient.
Successfully added 'Microsoft.WindowsAzure.Management.HDInsight 0.9.4951.25594' to HadoopClient.
```

■ **Note**   The version numbers that you see might change as new versions of the SDK are released.

You will find that the references to the respective .dll files have been added to your solution, as shown in Figure 4-4.

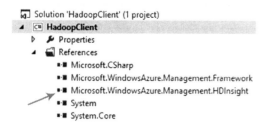

Figure 4-4. *The HadoopClient solution*

Connecting to Your Subscription

The first step towards consuming your Azure services from any client application is to upload a management certificate to Azure. This certificate will be subsequently used by the client applications to validate themselves while connecting to and using the Azure services. For more information about how to create and upload a management certificate, see the "Create a Certificate" section at the following link:

```
http://msdn.microsoft.com/en-us/library/windowsazure/gg981929.aspx
```

The HDInsight management package (`Microsoft.WindowsAzure.Management.HDInsight`) provides you with the .NET APIs to automate operations such as creating a cluster, creating a list and dropping existing clusters. The first thing that needs to be done, however, is providing the client applications with your Azure subscription certificate and its thumbprint. The standard .NET *X509* set of classes can be used to query the Azure certificate store. But before that, you will need to generate a unique thumbprint for your development system and bind it with your Azure subscription.

So the next task is to use Windows Azure PowerShell to bind your Azure subscription details to your development machine. You can install Azure PowerShell using the Web Platform Installer from the following link:

```
http://go.microsoft.com/fwlink/p/?linkid=320376&clcid=0x409
```

Accept the license agreement and you should see the installation screen for Azure PowerShell, as shown in Figure 4-5.

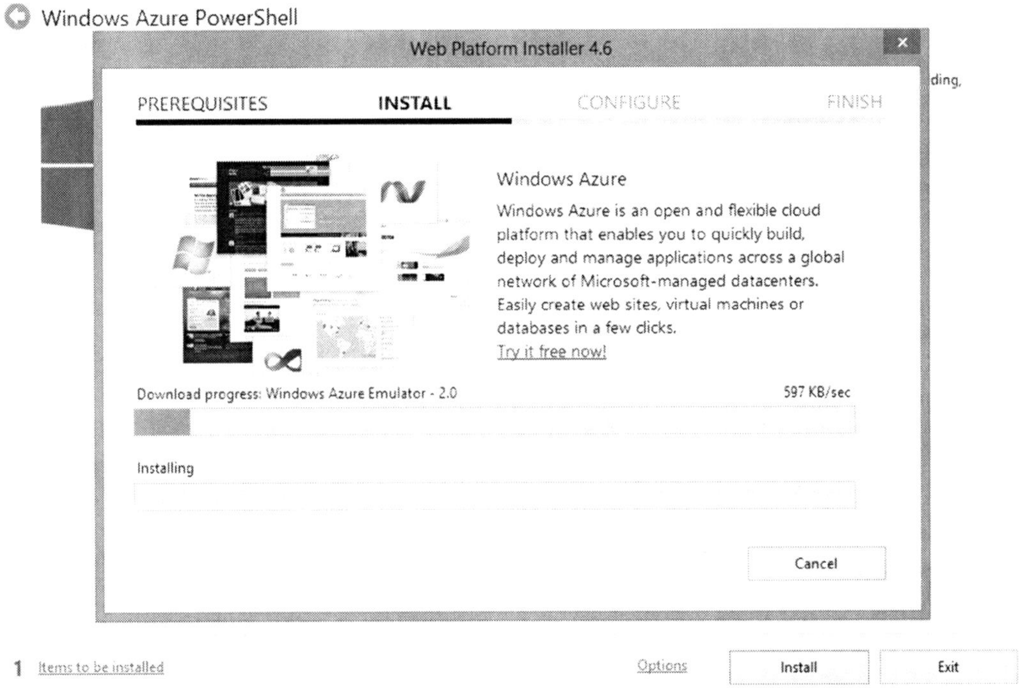

Figure 4-5. *Web Platform Installer*

Once the installation is complete, open the Windows Azure PowerShell console and execute the following command:

```
Get-AzurePublishSettingsFile
```

When prompted, download and save the publishing profile and note the path and name of the .publishsettings file. Then execute the following command to import the subscription with the proper path to the .publishsettings file:

```
Import-AzurePublishSettingsFile
C:\Users\<UserProfile>\Downloads\<SubscriptionName>-credentials.publishsettings
```

You should see a message in the PowerShell prompt about setting your default subscription. The message will be similar to the following:

```
VERBOSE:Setting: <subscription_name> as the default and current subscription. To view other
subscriptions use Get-AzureSubscription
```

Next, execute the Get-AzureSubscription command to list your subscription details as shown next. Note the thumbprint that is generated—you will be using this thumbprint further throughout your .NET solution:

```
PS C:\> Get-AzureSubscription

SubscriptionName             : <subscription_name>
SubscriptionId               : <subscription_Id>
ServiceEndpoint              : https://management.core.windows.net/
ActiveDirectoryEndpoint      :
ActiveDirectoryTenantId      :
IsDefault                    : True
Certificate                  : [Subject]
                        CN=Windows Azure Tools
 [Issuer]
                CN=Windows Azure Tools
[Serial Number]
                        793EE9285FF3D4A84F4F6B73994F3696
[Not Before]
                        12/4/2013 11:45:00 PM
[Not After]
                        12/4/2014 11:45:00 PM
[Thumbprint] <Thumbprint>
CurrentStorageAccountName    :
CurrentCloudStorageAccount   :
ActiveDirectoryUserId        :
```

Once this is done, you are ready to code your Visual Studio application.

■ **Note**   The .publishsettings file contains sensitive information about your subscription and credentials. Care should be taken to prevent unauthorized access to this file. It is highly recommended that you delete this file once it is imported successfully into PowerShell.

Coding the Application

In your HadoopClient solution, add a new class to your project and name it Constants.cs. There will be some constant values, such as the subscriptionID, certificate thumbprint, user names, passwords, and so on. Instead of writing them again and again, we are going to club these values in this class and refer to them from our program. Listing 4-1 shows the code in the Constants.cs file.

Listing 4-1. The Constants.cs File

```
using System;
using System.Collections.Generic;
using System.Linq;
using System.Text;

namespace HadoopClient
{
    public class Constants
    {
        public static Uri azureClusterUri = new Uri("https://democluster.azurehdinsight.net:443");
        public static string thumbprint = "your_subscription thumbprint";
        public static Guid subscriptionId = new Guid("your_subscription_id");
        public static string clusterUser = "admin";
        public static string hadoopUser = "hdp";
        public static string clusterPassword = "your_password";
        public static string storageAccount = "democluster.blob.core.windows.net";
        public static string storageAccountKey = "your_storage_key;
        public static string container = "democlustercontainer";
        public static string wasbPath =
 "wasb://democlustercontainer@democluster.blob.core.windows.net";
    }
}
```

When you choose your password, make sure to meet the following password requirements to avoid getting an error when you execute your program:

- The field must contain at least 10 characters.

- The field cannot contain the user name.

- The field must contain one each of the following: an uppercase letter, a lowercase letter, a number, a special character.

Next, navigate to the *Program.cs* file in the solution that has the Main() function, the entry point of a console application. You need to add the required references to access the certificate store for the Azure certificate as well as different HDInsight management operations. Go ahead and add the following using statements at the top of your *Program.cs* file:

```
using System.Security.Cryptography.X509Certificates;
using Microsoft.WindowsAzure.Management.HDInsight;
```

Create a new public function called ListClusters(). This function will have the code to query the certificate store and list the existing HDInsight clusters under that subscription. Listing 4-2 outlines the code for the ListClusters() function.

Listing 4-2. Enumerating Clusters in Your Subscription

```
public static void ListClusters()
{
    var store = new X509Store();
    store.Open(OpenFlags.ReadOnly);
    var cert = store.Certificates.Cast<X509Certificate2>()
```

```
.First(item => item.Thumbprint == Constants.thumbprint);
    var creds = new HDInsightCertificateCredential(Constants.subscriptionId, cert);
    var client = HDInsightClient.Connect(creds);
    var clusters = client.ListClusters();
    foreach (var item in clusters)
    {
Console.WriteLine("Cluster: {0}, Nodes: {1}", item.Name, item.ClusterSizeInNodes);
    }
}
```

Following are the first two lines of code. They connect to the X509 certificate store in read-only mode.

```
var store = new X509Store();
store.Open(OpenFlags.ReadOnly);
```

Next is a statement to load the Azure certificate based on the thumbprint:

```
        var cert = store.Certificates.Cast<X509Certificate2>().First(item =>
item.Thumbprint == Constants.thumbprint);
```

After loading the certificate, our next step is to create a client object based on the credentials obtained from the subscription ID and the certificate. We do that using the following statement:

```
    var creds = new HDInsightCertificateCredential(Constants.subscriptionId, cert);
    var client = HDInsightClient.Connect(creds);
```

Then we enumerate the HDInsight clusters under the subscription. The following lines grab the cluster collection and loops through each item in the collection:

```
        var clusters = client.ListClusters();
        foreach (var item in clusters)
        {
            Console.WriteLine("Cluster: {0}, Nodes: {1}",
item.Name, item.ClusterSizeInNodes);
        }
```

The WriteLine call within the loop prints the name of each cluster and its respective nodes.

You can run this code to list out your existing clusters in a console window. You need to add a call to this ListClusters() function in your Main() method and run the application. Because I have a couple of clusters deployed, I see the output shown in Figure 4-6 when I execute the preceding code.

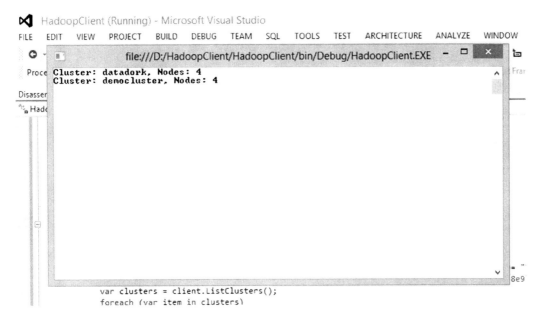

Figure 4-6. *The ListClusters( ) method*

You can use the `CreateCluster()` method of the SDK to programmatically deploy your HDInsight cluster. You will need to provide few mandatory parameters such as cluster name, location, storage account, and so on while calling the `CreateCluster()` method. Listing 4-3 contains the code block to provision a new cluster with two data nodes through .NET code.

Listing 4-3. The CreateCluster( ) Method

```
public static void CreateCluster()
{
    var store = new X509Store();
    store.Open(OpenFlags.ReadOnly);
    var cert = store.Certificates.Cast<X509Certificate2>()
.First(item => item.Thumbprint == Constants.thumbprint);
    var creds = new HDInsightCertificateCredential(Constants.subscriptionId, cert);
    var client = HDInsightClient.Connect(creds);

    //Cluster information
    var clusterInfo = new ClusterCreateParameters()
    {
Name = "AutomatedHDICluster",
Location = "East US",
DefaultStorageAccountName = Constants.storageAccount,
DefaultStorageAccountKey = Constants.storageAccountKey,
DefaultStorageContainer = Constants.container,
```

```
UserName = Constants.clusterUser,
Password = Constants.clusterPassword,
ClusterSizeInNodes = 2
    };
  var clusterDetails = client.CreateCluster(clusterInfo);
  ListClusters();
}
```

When you execute this method by similarly adding a call in Main(), you will see that a new cluster deployment has started in the Windows Azure Management Portal, as shown in Figure 4-7.

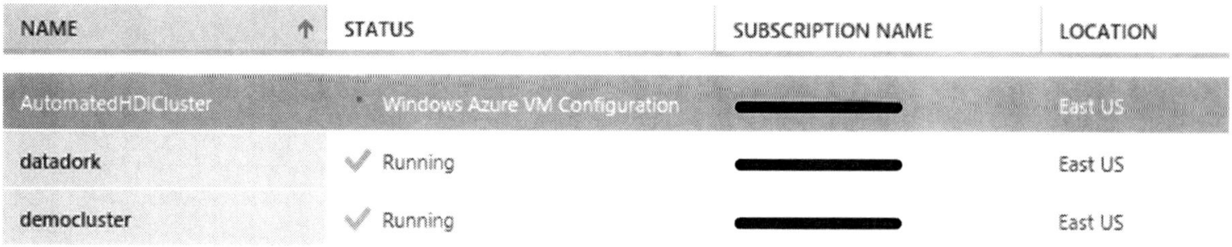

NAME	↑	STATUS	SUBSCRIPTION NAME	LOCATION
AutomatedHDICluster	•	Windows Azure VM Configuration	▬▬▬▬▬	East US
datadork	✓	Running	▬▬▬▬▬	East US
democluster	✓	Running	▬▬▬▬▬	East US

Figure 4-7. *New cluster provisioning*

Once the virtual machines are configured and the cluster creation is complete, you will see the cluster URL in your console application output. For example:

```
Created cluster: https://AutomatedHDICluster.azurehdinsight.net
```

You can call the ShowClusters() method again, and this time it will display three HDInsight clusters along with the new one just deployed:

```
Cluster: AutomatedHDICluster, Nodes: 2
Cluster: datadork, Nodes: 4
Cluster: democluster, Nodes: 4
```

You can also drop a cluster using the DeleteCluster() method of the .NET SDK. The code snippet in Listing 4-4 shows how to call the DeleteCluster() function.

Listing 4-4. The DeleteCluster( ) Method

```
public static void DeleteCluster()
{
    var store = new X509Store();
    store.Open(OpenFlags.ReadOnly);
    var cert = store.Certificates.Cast<X509Certificate2>().
        First(item => item.Thumbprint == Constants.thumbprint);
    var creds = new HDInsightCertificateCredential(Constants.subscriptionId, cert);
    var client = HDInsightClient.Connect(creds);
    client.DeleteCluster("AutomatedHDICluster");
    ListClusters();
}
```

After executing the `DeleteCluster()` method, you can go back to the Azure portal and confirm that the *AutomatedHDICluster*, which we just provisioned through code, no longer exists. You see only the two clusters that were previously created, as shown in Figure 4-8.

NAME	STATUS	SUBSCRIPTION NAME	LOCATION
datadork →	✔ Running	▬▬▬▬▬▬▬▬▬	East US
democluster	✔ Running	▬▬▬▬▬▬▬▬▬	East US

Figure 4-8. *AutomatedHDICluster is deleted*

Using the HDInsight management package, you can easily list, create, and delete your HDInsight clusters on Azure. Add a call to the functions we added earlier inside the `Main()` method, and call them sequentially to view the output in the console window. The complete code listing for the `Program.cs` file along with the `Main()` method is provided in Listing 4-5.

Listing 4-5. The Complete Code

```
using System;
using System.Collections.Generic;
using System.Linq;
using System.Text;
using System.Security.Cryptography.X509Certificates;
using Microsoft.WindowsAzure.Management.HDInsight;

namespace HadoopClient
{
    class Program
    {
        static void Main(string[] args)
        {
            ListClusters();
            CreateCluster();
            DeleteCluster();
            Console.ReadKey();
        }

        public static void ListClusters()
        {
            var store = new X509Store();
            store.Open(OpenFlags.ReadOnly);
            var cert = store.Certificates.Cast<X509Certificate2>().
First(item => item.Thumbprint == Constants.thumbprint);
            var creds = new HDInsightCertificateCredential(Constants.subscriptionId, cert);
            var client = HDInsightClient.Connect(creds);
            var clusters = client.ListClusters();
            foreach (var item in clusters)
```

```
        {
            Console.WriteLine("Cluster: {0}, Nodes: {1}", item.Name, item.ClusterSizeInNodes);
        }
    }

    public static void CreateCluster()
    {
        var store = new X509Store();
        store.Open(OpenFlags.ReadOnly);
        var cert = store.Certificates.Cast<X509Certificate2>().
First(item => item.Thumbprint == Constants.thumbprint);
        var creds = new HDInsightCertificateCredential(Constants.subscriptionId, cert);
        var client = HDInsightClient.Connect(creds);

        //Cluster information
        var clusterInfo = new ClusterCreateParameters()
        {
            Name = "AutomatedHDICluster",
            Location = "East US",
            DefaultStorageAccountName = Constants.storageAccount,
            DefaultStorageAccountKey = Constants.storageAccountKey,
            DefaultStorageContainer = Constants.container,
            UserName = Constants.clusterUser,
            Password = Constants.clusterPassword,
            ClusterSizeInNodes = 2
        };
        var clusterDetails = client.CreateCluster(clusterInfo);
        console.WriteLine("Cluster Created");
        ListClusters();
    }

    public static void DeleteCluster()
    {
        var store = new X509Store();
        store.Open(OpenFlags.ReadOnly);
        var cert = store.Certificates.Cast<X509Certificate2>().
First(item => item.Thumbprint == Constants.thumbprint);
        var creds = new HDInsightCertificateCredential(Constants.subscriptionId, cert);
        var client = HDInsightClient.Connect(creds);
        client.DeleteCluster("AutomatedHDICluster");
        console.WriteLine("Cluster Deleted");
        ListClusters();
    }

}
}
```

Windows Azure also exposes a set of PowerShell cmdlets for HDInsight to automate cluster management and job submissions. You can consider cmdlets as prebuilt PowerShell scripts that do specific tasks for you. The next section describes the PowerShell cmdlets for HDInsight for cluster provisioning.

Using the PowerShell cmdlets for HDInsight

The first step is to install the PowerShell cmdlets for HDInsight from the following URL:

```
http://www.microsoft.com/en-sg/download/details.aspx?id=40724
```

When prompted, save and unzip the zip files to a location of your choice. In my case, I chose my Visual Studio solution folder, as shown in Figure 4-9.

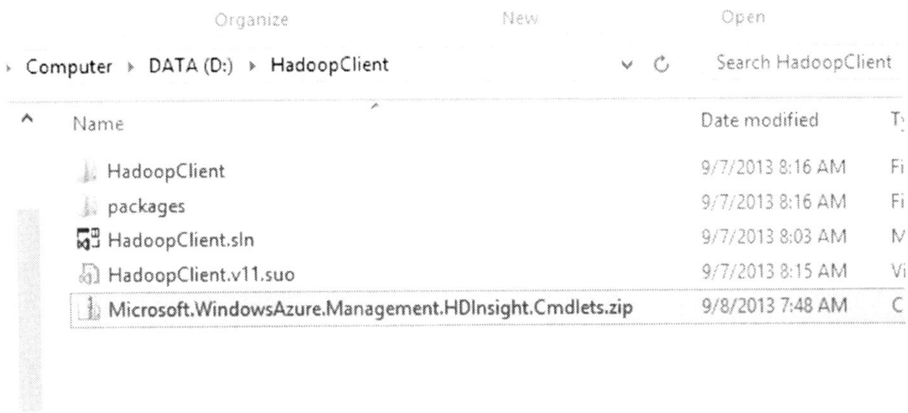

Figure 4-9. *HDInsight management cmdlets*

Note This step of installing the cmdlets won't be needed in the future when the HDInsight cmdlets are integrated and installed as part of Windows Azure PowerShell version 0.7.2. This book is based on Windows Azure PowerShell version 0.7.1, which does require this installation step.

Launch the Windows Azure PowerShell command prompt, and load the HDInsight cmdlet by executing the following command:

```
Import-Module "D:\HadoopClient\Microsoft.WindowsAzure.Management.HDInsight.Cmdlet.dll"
```

This will load the required set of HDInsightcmdlets in PowerShell:

```
PS C:\> Import-Module "D:\HadoopClient\Microsoft.WindowsAzure.Management.HDInsight.Cmdlet.dll"
VERBOSE: Loading module from path
'D:\HadoopClient\Microsoft.WindowsAzure.Management.HDInsight.Cmdlet.dll'.
VERBOSE: Importing cmdlet 'Add-AzureHDInsightMetastore'.
VERBOSE: Importing cmdlet 'Add-AzureHDInsightStorage'.
VERBOSE: Importing cmdlet 'New-AzureHDInsightCluster'.
VERBOSE: Importing cmdlet 'New-AzureHDInsightConfig'.
VERBOSE: Importing cmdlet 'Remove-AzureHDInsightCluster'.
VERBOSE: Importing cmdlet 'Get-AzureHDInsightCluster'.
VERBOSE: Importing cmdlet 'Set-AzureHDInsightDefaultStorage'.
```

■ **Note**   The path of the `Microsoft.WindowsAzure.Management.HDInsight.Cmdlet.dll` file might vary depending on where you choose to download it.

In some cases, depending on your operating system and account security policies, you might need to unblock the downloaded cmdlets.zip file to let it load into PowerShell. You can do it from the properties of the .zip file, as shown in Figure 4-10.

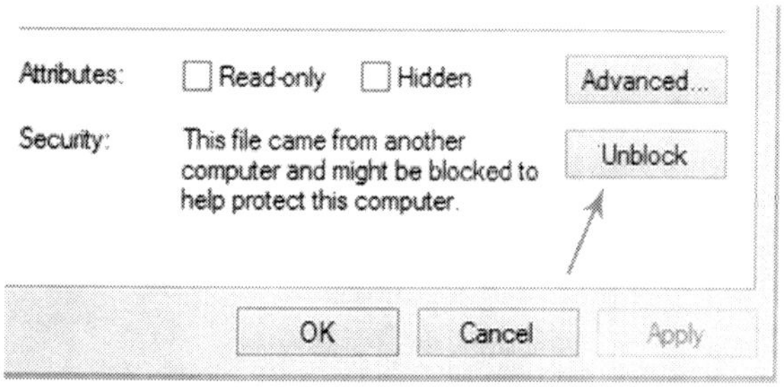

Figure 4-10. Unblock downloaded content

Also, depending on your system's security configuration, you might need to set PowerShell execution policy so that it can execute remotely-signed assemblies. To do this, launch Windows Azure PowerShell as an administrator and execute the following command:

```
Set-ExecutionPolicy RemoteSigned
```

If you do not do this, and your security setting does not allow you to load a .dll file that is built and signed on a remote system, you will see similar error messages in PowerShell while trying to import the `Microsoft.WindowsAzure.Management.HDInsight.Cmdlet.dll`:

```
Import-Module : The specified module
'D:\Microsoft.WindowsAzure.Management.HDInsight.Cmdlets\Microsoft.WindowsAzure.Management.
HDInsight.Cmdlet.dll' wasnot loaded because no valid module file was found in any module directory.
```

Once the cmdlet is successfully loaded, the first thing you need to do is associate the subscription id and the management certificate for your Azure subscription with the cmdlet variables. You can use the following commands to set them:

```
$subid = Get-AzureSubscription -Current | %{ $_.SubscriptionId }
$cert = Get-AzureSubscription -Current  | %{ $_.Certificate }
```

Once they are set, you can execute the following command to list your existing HDInsight clusters:

```
Get-AzureHDInsightCluster -SubscriptionId $subid -Certificate $cert
```

Because I have two clusters, I get the following output:

```
PS C:\> Get-AzureHDInsightCluster -SubscriptionId $subid -Certificate $cert

Name               : datadork
ConnectionUrl      : https://datadork.azurehdinsight.net
State              : Running
CreateDate         : 8/16/2013 9:19:09 PM
UserName           : admin
Location           : East US
ClusterSizeInNodes : 4

Name               : democluster
ConnectionUrl      : https://democluster.azurehdinsight.net
State              : Running
CreateDate         : 6/26/2013 6:59:30 PM
UserName           : admin
Location           : East US
ClusterSizeInNodes : 4
```

To provision a cluster, you need to specify a storage account. The HDInsight cmdlets will need to get the key for your storage account dedicated to the cluster. If you remember, I am using my storage account, called hdinsightstorage, for all my clusters. Issuing the following PowerShell command will populate the cmdlet variable with the storage account key:

```
$key1 = Get-AzureStorageKey hdinsightstorage | %{ $_.Primary }
```

On successful access to the storage account key, you will see messages similar to the following ones:

```
PS C:\> $key1 = Get-AzureStorageKey hdinsightstorage | %{ $_.Primary }
VERBOSE: 8:50:29 AM - Begin Operation: Get-AzureStorageKey
VERBOSE: 8:50:34 AM - Completed Operation: Get-AzureStorageKey
```

If you provide the wrong storage account name or one that belongs to a different subscription, you might get error messages like the following ones while trying to acquire the storage account key:

```
PS C:\> $key1 = Get-AzureStorageKey hdinsightstorage | %{ $_.Primary }
VERBOSE: 1:30:18 PM - Begin Operation: Get-AzureStorageKey
Get-AzureStorageKey : "An exception occurred when calling the ServiceManagement API. HTTP Status
Code: 404.

ServiceManagement Error Code: ResourceNotFound.
Message: The storage account 'hdinsightstorage' was not found.. OperationTracking
ID:72c0c6bb12b94f849aa8884154655089."
```

■ **Note** If you have multiple subscriptions, you can use Set-AzureSubscription -DefaultSubscription "<Your_Subscription_Name>" to set to default subscription in PowerShell where your cluster storage accounts reside.

Now you have all the necessary information to spin up a new cluster using the cmdlet. The following snippet shows you the command with all the required parameters to provision a new HDInsight cluster:

```
New-AzureHDInsightCluster -SubscriptionId $subid -Certificate $cert -Name AutomatedHDI
-Location "East US" - DefaultStorageAccountName hdinsightstorage.blob.core.windows.net
-DefaultStorageAccountKey $key1 -DefaultStorageContainerName "democluster"
-UserName "admin" -Password "***************" -ClusterSizeInNodes 2
```

Your Windows Azure Management Portal will soon display the progress of your cluster provisioning, as shown in Figure 4-11.

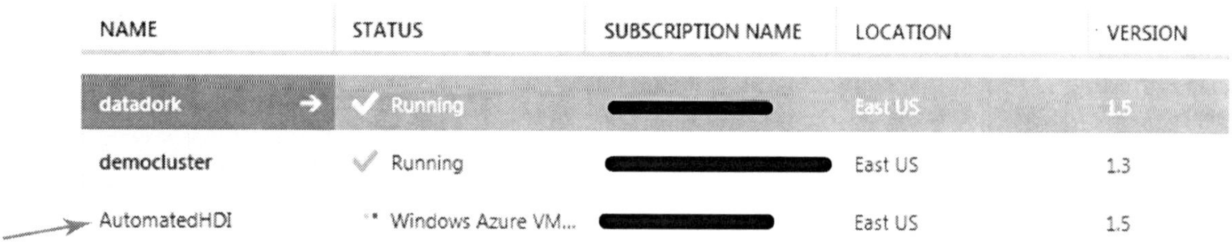

NAME	STATUS	SUBSCRIPTION NAME	LOCATION	VERSION
datadork	✔ Running	▬▬▬▬▬	East US	1.5
democluster	✔ Running	▬▬▬▬▬	East US	1.3
AutomatedHDI	·* Windows Azure VM...	▬▬▬▬▬	East US	1.5

Figure 4-11. *Cluster provisioning in progress*

On completion of the cluster creation, you will see the PowerShell prompt displaying the details of the newly created cluster:

```
PS C:\> New-AzureHDInsightCluster
-SubscriptionId $subid
-Certificate $cert
-Name AutomatedHDI
-Location "East US"
-DefaultStorageAccountName hdinsightstorage.blob.core.windows.net
-DefaultStorageAccountKey $key1
-DefaultStorageContainerName "democluster"
-UserName "admin"
-Password "******************"
-ClusterSizeInNodes 2

Name                : AutomatedHDI
ConnectionUrl       : https://AutomatedHDI.azurehdinsight.net
State               : Running
CreateDate          : 9/8/2013 3:34:07 AM
UserName            : admin
Location            : East US
ClusterSizeInNodes  : 2
```

If there is an error in the specified command, the PowerShell console will show you the error messages. For example, if the supplied cluster password does not meet the password-compliance policy, you will see an error message similar to the following while trying to provision a new cluster:

```
New-AzureHDInsightCluster : Unable to complete the 'Create' operation. Operation failed with code '400'.
Cluster leftbehind state: 'Specified Cluster password is invalid. Ensure password is 10 characters long
and has atleast onenumber, one uppercase and one special character(spaces not allowed)'. Message: 'NULL'.
```

To delete the newly provisioned cluster, you can use the `Remove-AzureHDInsightCluster` command as shown here:

```
Remove-AzureHDInsightCluster AutomatedHDI -SubscriptionId $subid -Certificate $cert
```

Table 4-2 summarizes the commands available in the HDInsight cmdlet and provides a brief overview of their functions.

Table 4-2. *HDInsight cmdlet commands*

Command	Function
Add-AzureHDInsightMetastore	Customize the Hive/Oozie metadata storage location.
Add-AzureHDInsightStorage	Add a new storage account to the subscription.
New-AzureHDInsightCluster	Provision a new HDInsight cluster.
New-AzureHDInsightConfig	Used to parameterize HDInsight cluster properties like number of nodes based on configured values.
Remove-AzureHDInsightCluster	Delete anHDInsight cluster.
Get-AzureHDInsightCluster	List the provisioned HDInsight cluster for the subscription.
Set-AzureHDInsightDefaultStorage	Set the default storage account for HDInsight cluster creations.

PowerShellcmdlets give you the flexibility of really taking advantage of the elasticity of services that Azure HDInsight provides. You can create a PowerShell script that will spin up your Hadoop cluster when required, submit jobs for processing, and shut the cluster down once the output is written back into Azure blob storage. This process is possible because the storage used for input and output is Azure blob storage. As such, the cluster is needed only for compute operations and not storage. During the creation, the cluster name and number of hosts can be specified, and during the job submission, the input and output paths can be specified as well. One could, of course, customize these scripts to include additional parameters such as the number of mappers, additional job arguments, and so on.

■ **Note** Microsoft consultant Carl Nolan has a wonderful blog about using PowerShell cmdlets to provide a mechanism for managing an elastic service. You can read his blog at `http://blogs.msdn.com/b/carlnol/archive/2013/06/07/managing-your-hdinsight-cluster-with-powershell.aspx`.

Command-Line Interface (CLI)

The command line is an open source, cross-platform interface for managing HDInsight clusters. It is implemented in Node.js. Thus, it is usable from multiple platforms, such as Windows, Mac, Linux, and so on. The source code is available at the GitHub web site:

```
https://github.com/WindowsAzure/azure-sdk-tools-xplat
```

The sequence of operations in CLI is pretty much the same as in PowerShell. You have to download and import the Azure `.publishsettings` file as a persistent local config setting that the command-line interface will use for its subsequent operations.

The CLI can be installed in one of two ways:

- From the Node.js Package Manager (NPM), do the following:

    a. Navigate to www.nodejs.org.

    b. Click on Install, and follow the instructions, accepting the default settings.

    c. Open a command prompt, and execute the following command:
        ```
        npm install -g azure-cli
        ```

- From the Windows Installer, do the following:

    a. Navigate to http://www.windowsazure.com/en-us/downloads/.

    b. Scroll down to the *Command line tools* section, and then click *Cross-platform Command Line Interface* and follow the Web Platform Installer wizard instructions.

Once the installation is complete, you need to verify the installation. To do that, open Windows Azure Command Prompt and execute the following command:

```
azure hdinsight -h
```

If the installation is successful, this command should display the help regarding all the HDInsight commands that are available in CLI.

■ **Note**   If you get an error that the command is not found, make sure you have the path C:\Program Files (x86)\ Microsoft SDKs\Windows Azure\CLI\wbin\ to the PATH environment variable in the case of Windows Installer. For NPM, make sure that the path C:\Program Files (x86)\nodejs;C:\Users\[username]\AppData\Roaming\npm\ is appended to the PATH variable.

Once it is installed, execute the following command to download and save the publishsettings file:

```
azure account download
```

You should see output similar to the following once the file is downloaded:

```
info:    Executing command account download
info:    Launching browser to http://go.microsoft.com/fwlink/?LinkId=254432
help:    Save the downloaded file, then execute the command
help:    account import <file>
info:    account download command OK
```

The next step is to import the file in CLI using the following command:

```
azure account import<publishsettings file>
```

The file should be successfully imported, and the output will be similar to the following:

```
info:    Executing command account import
info:    Found subscription: <subscription_name>
info:    Setting default subscription to: <subscription_name>
```

```
info:    Use "azure account set" to change to a different one
info:    Setting service endpoint to: https://management.core.windows.net/
warn:    The "file_name" file contains sensitive information
warn:    Remember to delete it now that it has been imported
info:    Account publish settings imported successfully
info:    account import command OK
```

To list the existing HDInsight clusters in your subscription, you can use the following command:

```
azure hdinsight cluster list
```

The output will be the list of your already-provisioned clusters in the running state . It will be something similar to the following, which I generated with four HDInsight clusters under my subscription:

```
info:    Executing command hdinsight cluster list
+ Getting HDInsight servers
data:    Name          Location       State
data:    -----------   ------------   -------
data:    SDPHDI1       East US        Running
data:    democluster   North Europe   Running
data:    datadork      West US        Running
data:    tutorial      West US        Running
info:    hdinsight cluster list command OK
```

You can use the azure hdinsight cluster delete <ClusterName> command to delete any existing cluster. To create a new cluster using the CLI, you need to provide the cluster name, subscription information, and other details, similar to provisioning a cluster using PowerShell or the .NET SDK. Listing 4-6 shows a sample command to create a new HDInsight cluster using CLI.

Listing 4-6. Creating a Cluster Using CLI

```
azure hdinsight cluster create --clusterName <ClusterName>
--storageAccountName <StorageAccountName>
--storageAccountKey <storageAccountKey>
--storageContainer <StorageContainer>
--nodes <NumberOfNodes>
--location <DataCenterLocation>
--username <HDInsightClusterUsername>
--clusterPassword <HDInsightClusterPassword>
```

Typically, you provision an HDInsight cluster, run jobs on it, and then delete the cluster to cut down the cost. The command-line interface also gives you the option to save the configurations into a file so that you can reuse it every time you provision a cluster. This is basically another way of automating cluster provisioning and several other administrative tasks. For comprehensive reference documentation on the cross-platform CLI tools, have a look at the following:

```
http://www.windowsazure.com/en-us/manage/install-and-configure-cli/
```

Summary

The Windows Azure HDInsight service exposes a set of .NET-based interfaces to control your clusters programmatically. While .NET languages like C# are a popular choice with many skilled developers, HDInsight also has a tight coupling with Windows Azure PowerShell and provides a set of useful cmdlets for cluster management. PowerShell is a common choice of Windows administrators for creating a script-based management infrastructure. The combination of the .NET SDK and PowerShell provide an automated way of implementing on-demand cluster provisioning and job submission, thus leveraging the full flexibility of Azure elastic services. In addition to these .NET APIs and PowerShell cmdlets, there is also a multiplatform-aware, node.js-based, command-line interface that can be used for cluster management programmatically.

Because storage is isolated and retained in Azure blobs, you no longer need to have your Hadoop clusters online and pay for computation hours. In this chapter, you saw how to use the .NET APIs, PowerShell, and cross-platform CLI commands for basic cluster-management operations. Currently, the Hadoop .NET SDK provides API access to aspects of HDInsight, including HDFS, HCatalog, Oozie, and Ambari. There are also libraries for MapReduce and LINQ to Hive. The latter is really interesting because it builds on the established technology for .NET developers to access most data sources to deliver the capabilities of the de facto standard for Hadoop data query.

CHAPTER 5

Submitting Jobs to Your HDInsight Cluster

Apart from the cluster-management operations you saw in the previous chapter, you can use the .NET SDK and the Windows PowerShell cmdlets to control your job submission and execution in your HDInsight cluster. The jobs are typically MapReduce jobs because that is the only thing that Hadoop understands. You can write your MapReduce jobs in .NET and also use supporting projects—such as Hive, Pig, and so forth—to avoid coding MapReduce programs, which can often be tedious and time consuming.

In all the samples I have shown so far, I used the command-line consoles. However, this does not need to be the case; you can also use PowerShell. The Console application that is used to submit the MapReduce jobs calls a .NET Submissions API. As such, one can call the .NET API directly from within PowerShell, similar to the cluster-management operations. You will use the same console application you created in the previous chapter and add the functions for job submissions. In this chapter, you will learn how to implement a custom MapReduce program in .NET and execute it as a Hadoop job. You will also take a look at how to execute the sample wordcount MapReduce job and a Hive query using .NET and PowerShell.

Using the Hadoop .NET SDK

Hadoop streaming is a utility that comes with the Hadoop distribution. The utility allows you to create and run MapReduce jobs with any executable or script as the mapper and/or the reducer. This is essentially a Hadoop API to MapReduce that allows you to write map and reduce functions in languages other than Java (.NET, Perl, Python, and so on). Hadoop Streaming uses Windows streams as the interface between Hadoop and the program, so you can use any language that can read standard input and write to standard output to write the MapReduce program. This functionality makes streaming naturally suited for text processing. In this chapter, I focus only on .NET to leverage Hadoop streaming.

The mapper and reducer parameters are .NET types that derive from base Map and Reduce abstract classes. The input, output, and files options are analogous to the standard Hadoop streaming submissions. The mapper and reducer allow you to define a .NET type derived from the appropriate abstract base classes.

The objective in defining these base classes was not only to support creating .NET Mapper and Reducer classes but also to provide a means for *Setup* and *Cleanup* operations to support in-place Mapper/Combiner/Reducer optimizations, utilize IEnumerable and sequences for publishing data from all classes, and finally provide a simple submission mechanism analogous to submitting Java-based jobs.

The basic logic behind MapReduce is that the Hadoop text input is processed and each input line is passed into the *Map* function, which parses and filters the key/value pair for the data. The values are then sorted and merged by Hadoop. The processed mapped data is then passed into the *Reduce* function, as a key and corresponding sequence of strings, which then defines the optional output value. One important thing to keep in mind is that Hadoop Streaming is based on text data. Thus, the inputs into the MapReduce are strings or *UTF8*-encoded bytes. However, when you are performing the MapReduce operations, strings are not always suitable, but the operations do need to be able to be represented as strings.

Adding the References

Open the C# console application HadoopClient that you created in the previous chapter. Once the solution is opened, open the NuGet *Package Manager Console* and import the MapReduce NuGet package by running the following command:

```
install-package Microsoft.Hadoop.MapReduce
```

This should import the required .dll, along with any dependencies it may have. You will see output similar to the following:

```
PM> install-package Microsoft.Hadoop.MapReduce
Attempting to resolve dependency 'Newtonsoft.Json (≥ 4.5.11)'.
Installing 'Newtonsoft.Json 4.5.11'.
Successfully installed 'Newtonsoft.Json 4.5.11'.
Installing 'Microsoft.Hadoop.MapReduce 0.9.4951.25594'.
Successfully installed 'Microsoft.Hadoop.MapReduce 0.9.4951.25594'.
Adding 'Newtonsoft.Json 4.5.11' to HadoopClient.
Successfully added 'Newtonsoft.Json 4.5.11' to HadoopClient.
Adding 'Microsoft.Hadoop.MapReduce 0.9.4951.25594' to HadoopClient.
Successfully added 'Microsoft.Hadoop.MapReduce 0.9.4951.25594' to HadoopClient.
Setting MRLib items CopyToOutputDirectory=true
```

▪ **Note**   The version numbers displayed while installing the NuGet package might change with future version updates of the SDK.

Once the NuGet package has been added, add a reference to the dll file in your code:

```
using Microsoft.Hadoop.MapReduce;
```

Once these required references are added, you are ready to code your MapReduce classes and job-submission logic in your application.

Submitting a Custom MapReduce Job

In the previous chapter, we already created the Constants.cs class to re-use several constant values, like your Azure cluster url, storage account, containers and so on. The code in the class file should look similar to Listing 5-1.

Listing 5-1. The Constants class

```
using System;
using System.Collections.Generic;
using System.Linq;
using System.Text;

namespace HadoopClient
{
    public class Constants
```

```
{
    public static Uri azureClusterUri
        = new Uri("https://democluster.azurehdinsight.net:443");
    public static string clusterName = "democluster";
    public static string thumbprint = "Your_Certificate_Thumbprint";
    public static Guid subscriptionId = new Guid("Your_Subscription_Id");
    public static string clusterUser = "admin";
    public static string hadoopUser = "hdp";
    public static string clusterPassword = "Your_Password";
    public static string storageAccount = "democluster";
    public static string storageAccountKey = "Your_storage_Key";
    public static string container = "democlustercontainer";
    public static string wasbPath =
"wasb://democlustercontainer@democluster.blob.core.windows.net";
    }
}
```

■ **Note**   Connection to the HDInsight cluster defaults to the standard Secure Sockets Layer (SSL) port 443. However, if you have a cluster prior to version 2.1, the connection is made through port 563.

The constant hadoopUser is the user account that runs the Hadoop services on the NameNode. By default, this user is hdp in an HDInsight distribution. You can always connect remotely to the NameNode and find this service account from the Windows Services console, as shown in Figure 5-1.

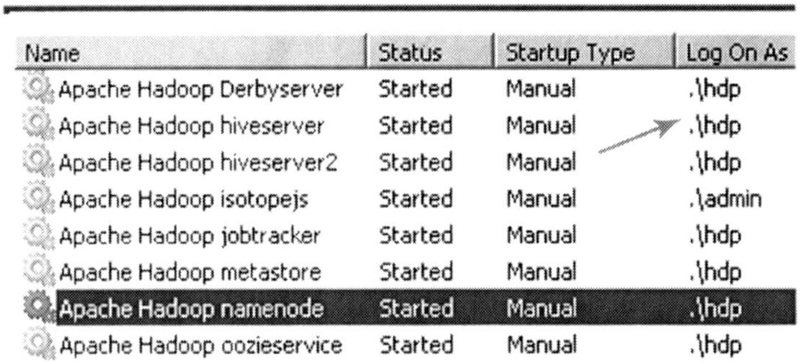

Name	Status	Startup Type	Log On As
Apache Hadoop Derbyserver	Started	Manual	.\hdp
Apache Hadoop hiveserver	Started	Manual	.\hdp
Apache Hadoop hiveserver2	Started	Manual	.\hdp
Apache Hadoop isotopejs	Started	Manual	.\admin
Apache Hadoop jobtracker	Started	Manual	.\hdp
Apache Hadoop metastore	Started	Manual	.\hdp
Apache Hadoop namenode	Started	Manual	.\hdp
Apache Hadoop oozieservice	Started	Manual	.\hdp

Figure 5-1. Hadoop service account

You will use these class variables henceforth in the different methods you call from your client applications. Using them helps to improve the readability as well as the management of the code.

Adding the MapReduce Classes

Hadoop Streaming is an interface for writing MapReduce jobs in the language of your choice. Hadoop SDK for .NET is a wrapper to Streaming that provides a convenient experience for .NET developers to develop MapReduce programs. The jobs can be submitted for execution via the API. The command is displayed on the JobTracker web interface and can be used for direct invocation if required.

A .NET map-reduce program comprises a number of parts (which are described in Table 5-1):

- Job definition
- Mapper, Reducer, and Combiner classes
- Input data
- Job executor

Table 5-1. *The function of .NET MapReduce components*

Component	Function
Job definition	This class has the declarations for Mapper, Reducer, and Combiner types as well as the job configuration settings.
Map, Reduce, and Combine	These are the actual classes you use to implement your processing logic.
Input data	The data for the MapReduce job to process.
Job executor	The entry point of your program—for example, the Main() method—which invokes the HadoopJobExecutor API.

In the following section, you will create a MapReduce program that calculates the square root of all the integer values provided as input and writes the output data to the file system.

In your HadoopClient solution, add three classes—SquareRootMapper, SquareRootReducer, and SquareRootJob—as shown in Figure 5-2.

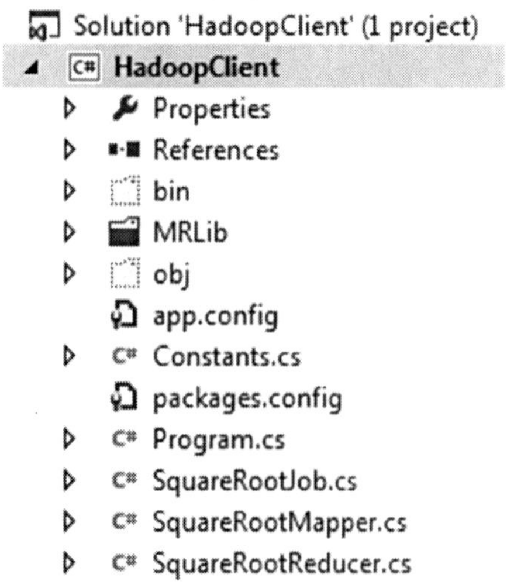

Figure 5-2. *Mapper, Reducer, and Job classes*

You need to inherit your mapper class from the .NET Framework base class, MapperBase, and override its Map() method. Listing 5-2 shows the code for the mapper class.

Listing 5-2. SquareRootMapper.cs

```
using System;
using System.Collections.Generic;
using System.Linq;
using System.Text;
using Microsoft.Hadoop.MapReduce;

namespace HadoopClient
{
    class SquareRootMapper: MapperBase
    {
        public override void Map(string inputLine, MapperContext context)
        {
            int input = int.Parse(inputLine);

            // Find the square root.
            double root = Math.Sqrt((double)input);

            // Write output.
            context.EmitKeyValue(input.ToString(), root.ToString());
        }
    }
}
```

The Map() function alone is enough for a simple calculation like determining square roots. So your Reducer class would not have any processing code or logic in this case. You can choose to omit it because Reduce and Combine are optional operations in a MapReduce job. However, it is a good practice to have the skeleton class for the Reducer, which derives from the ReducerCombinerBase .NET Framework class, as shown in Listing 5-3. You can write your code in the overridden Reduce() method later if you need to implement any reduce operations.

Listing 5-3. SquareRootReducer.cs

```
using System;
using System.Collections.Generic;
using System.Linq;
using System.Text;
using Microsoft.Hadoop.MapReduce;

namespace HadoopClient
{
    class SquareRootReducer: ReducerCombinerBase
    {
        public override void Reduce(string key, IEnumerable<string>
        values, ReducerCombinerContext context)
        {
            //throw new NotImplementedException();
        }
    }
}
```

■ **Note** Windows Azure MSDN documentation has a sample C# wordcount program that implements both the Mapper and Reducer classes: http://www.windowsazure.com/en-us/documentation/articles/hdinsight-sample-csharp-streaming/.

Once the Mapper and Reducer classes are defined, you need to implement the HadoopJob class. This consists of the configuration information for your job—for example, the input data and the output folder path. Listing 5-4 shows the code snippet for the SquareRootJob class implementation.

Listing 5-4. SquareRootJob.cs

```
using System;
using System.Collections.Generic;
using System.Linq;
using System.Text;
using Microsoft.Hadoop.MapReduce;

namespace HadoopClient
{
    class SquareRootJob: HadoopJob<SquareRootMapper>
    {
        public override HadoopJobConfiguration Configure(ExecutorContext context)
        {
            var config = new HadoopJobConfiguration
            {
                InputPath = Constants.wasbPath + "/example/data/Numbers.txt",
                OutputFolder = Constants.wasbPath + "/example/data/SqaureRootOutput"
            };
            return config;
        }
    }
}
```

■ **Note** I chose \example\data as the input path where I would have my source file, Numbers.txt. The output will be generated in the \example\data\SquareRootOutput folder. This output folder will be overwritten each time the job runs. If you want to preserve an existing job output folder, make sure to change the output folder name each time before job execution.

Per the configuration option specified in the job class, you need to upload the input file Numbers.txt and the job will write the output data to a folder called SquareRootOutput in Windows Azure Storage Blob (WASB). This will be the \example\data directory of the democlustercontainer in the democluster storage account as specified by the constant wasbPath in the Constants.cs class.

Running the MapReduce Job

Just before running the job, you need to upload the input file Numbers.txt to the storage account. Here is the content of a sample input file:

```
100
200
300
400
500
```

Use the PowerShell script shown in Listing 5-5 to upload the file to your blob container.

Listing 5-5. Using PowerShell to upload a file

```
$subscriptionName = "Your_Subscription_Name"
$storageAccountName = "democluster"
$containerName = "democlustercontainer"
#This path may vary depending on where you place the source .csv files.
$fileName ="C:\Numbers.txt"
$blobName = "\example\data\Numbers.txt"
# Get the storage account key
Select-AzureSubscription $subscriptionName
$storageaccountkey = get-azurestoragekey $storageAccountName | %{$_.Primary}
# Create the storage context object
$destContext = New-AzureStorageContext -StorageAccountName $storageAccountName -StorageAccountKey
$storageaccountkey
# Copy the file from local workstation to the Blob container
Set-AzureStorageBlobContent -File $fileName -Container $containerName
-Blob $blobName -context $destContext
```

On successful execution, you should see output similar to the following:

```
Container Uri:
https://democluster.blob.core.windows.net/democlustercontainer

Name          BlobType    Length      ContentType   LastModified  SnapshotTime
----          --------    ------      -----------   ------------  ------------
/example/d... BlockBlob   23          applicatio... 12/9/2013 ...
```

You can also verify that the file exists in your blob container through the Azure Management portal, as shown in Figure 5-3.

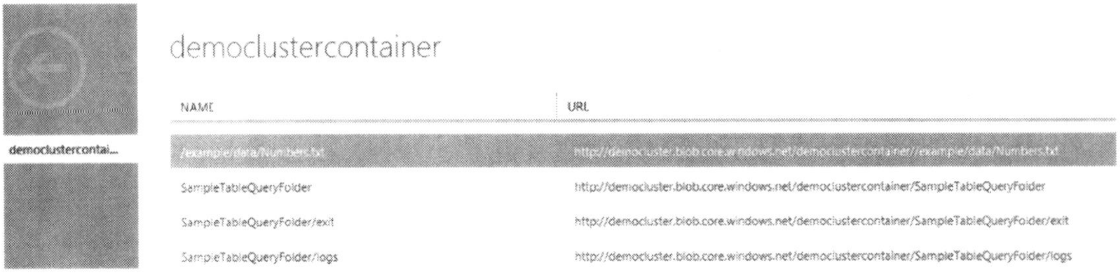

Figure 5-3. *Numbers.txt uploaded in blob*

You are now ready to invoke the job executor from your Main() method using the ExecuteJob() method. In your Program.cs file, add a function DoCustomMapReduce() with code like that in Listing 5-6. Note that this chapter will be using several inbuilt .NET Framework classes for IO, threading, and so on. Make sure you have the following set of using statements in your Program.cs file:

```
using System;
using System.Collections.Generic;
using System.Linq;
using System.Text;
using System.Security.Cryptography.X509Certificates;
using Microsoft.WindowsAzure.Management.HDInsight;
using Microsoft.Hadoop.MapReduce;
using Microsoft.Hadoop.Client;
//For Stream IO
using System.IO;
//For Ambari Monitoring Client
using Microsoft.Hadoop.WebClient.AmbariClient;
using Microsoft.Hadoop.WebClient.AmbariClient.Contracts;
//For Regex
using System.Text.RegularExpressions;
//For thread
using System.Threading;
//For Blob Storage
using Microsoft.WindowsAzure.Storage;
using Microsoft.WindowsAzure.Storage.Blob;
```

Listing 5-6. DoCustomMapReduce method

```
public static void DoCustomMapReduce()
{
            Console.WriteLine("Starting MapReduce job. Log in remotely to your Name Node and check
progress from JobTracker portal with the returned JobID...");
        IHadoop hadoop = Hadoop.Connect(Constants.azureClusterUri, Constants.clusterUser,
                        Constants.hadoopUser, Constants.clusterPassword,
                        Constants.storageAccount, Constants.storageAccountKey,
                        Constants.container, true);
        var output = hadoop.MapReduceJob.ExecuteJob<SquareRootJob>();
}
```

Finally, add a call to the DoCustomMapReduce() method from your Main() function. The Main() function in your Program.cs file should now look like Listing 5-7.

Listing 5-7. Main() method

```
static void Main(string[] args)
        {
            //ListClusters();
            //CreateCluster();
            //DeleteCluster();
            DoCustomMapReduce();
            Console.ReadKey();
        }
```

Execute the HadoopClient project and your console output should display messages similar to the following:

```
Starting MapReduce job. Log in remotely to your Name Node and check progress from JobTracker portal
with the returned JobID...

File dependencies to include with job:
[Auto-detected] D:\HadoopClient\HadoopClient\bin\Debug\HadoopClient.vshost.exe
[Auto-detected] D:\HadoopClient\HadoopClient\bin\Debug\HadoopClient.exe
[Auto-detected] D:\HadoopClient\HadoopClient\bin\Debug\Microsoft.Hadoop.MapReduce.dll
[Auto-detected] D:\HadoopClient\HadoopClient\bin\Debug\Microsoft.Hadoop.WebClient.dll
[Auto-detected] D:\HadoopClient\HadoopClient\bin\Debug\Newtonsoft.Json.dll
[Auto-detected] D:\HadoopClient\HadoopClient\bin\Debug\Microsoft.Hadoop.Client.dll
```

Job job_201309161139_003 completed.

⬚ **Note** I commented out the cluster management method calls in the Main() function because we are focusing on only the MapReduce job part. Also, you may see a message about deleting the output folder if it already exists.

If, for some reason, the required environment variables are not set, you may get an error like the following one while executing the project, which indicates the environment is not suitable:

```
Environment Vairable not set: HADOOP_HOME
Environment Vairable not set: Java_HOME
```

If you encounter such a situation, add the following two lines of code to set the variables at the top of your DoCustomMapReduce() method:

```
//This is constant
Environment.SetEnvironmentVariable("HADOOP_HOME", @"c:\hadoop");
//Needs to be Java path of the development machine
Environment.SetEnvironmentVariable("Java_HOME", @"c:\hadoop\jvm");
```

On successful completion, the job returns the job id. Using that, you can track the details of the job in the Hadoop MapReduce Status or JobTracker portal by remotely connecting to the NameNode. Figure 5-4 shows the preceding job's execution history in the JobTracker web application.

Running Jobs

Completed Jobs

Jobid	Started	Priority	User	Name	Map % Complete	Map Total	Maps Completed	Reduce % Complete	
job_201309161139_0001	Mon Sep 16 12:10:43 GMT 2013	NORMAL	admin	TempletonControllerJob	100.00%	1	1	100.00%	
job_201309161139_0002	Mon Sep 16 12:11:06 GMT 2013	NORMAL	admin	streamjob6737947396646342963.jar	100.00%	2	2	100.00%	
job_201309161139_0003	Mon Sep 16 12:11:51 GMT 2013	NORMAL	admin	TempletonControllerJob	100.00%	1	1	100.00%	

Figure 5-4. JobTracker portal

You can click on the job in the portal to further drill down into the details of the operation, as shown in Figure 5-5.

Job Setup: <u>Successful</u>
Status: Succeeded
Started at: Mon Sep 16 12:11:51 GMT 2013
Finished at: Mon Sep 16 12:12:37 GMT 2013
Finished in: 45sec
Job Cleanup: <u>Successful</u>
Job Scheduling information: 0 running map tasks using 0 map slots. 0 additional slots reserved. 0 running reduce tasks using 0 reduce slots.

Kind	% Complete	Num Tasks	Pending	Running	Complete	Killed	Failed/Killed Task Attempts
<u>map</u>	100.00%	1	0	0	<u>1</u>	0	0 / 0
<u>reduce</u>	100.00%	0	0	0	0	0	0 / 0

	Counter	Map	Reduce	Total
Job Counters	SLOTS_MILLIS_MAPS	0	0	40,297
	Total time spent by all reduces waiting after reserving slots (ms)	0	0	0
	Total time spent by all maps waiting after reserving slots (ms)	0	0	0
	Launched map tasks	0	0	1
	SLOTS_MILLIS_REDUCES	0	0	0
File Output Format Counters	Bytes Written	0	0	0
File Input Format Counters	Bytes Read	0	0	0
FileSystemCounters	FILE_BYTES_READ	1,064	0	1,064
	HDFS_BYTES_READ	45	0	45
	ASV_BYTES_WRITTEN	242	0	242
	FILE_BYTES_WRITTEN	28,669	0	28,669
Map-Reduce Framework	Map input records	0	0	0
	Physical memory (bytes) snapshot	163,479,552	0	163,479,552
	Spilled Records	0	0	0
	Total committed heap usage (bytes)	514,523,136	0	514,523,136
	CPU time spent (ms)	5,983	0	5,983
	Virtual memory (bytes) snapshot	653,365,248	0	653,365,248

Figure 5-5. *MapReduce job details*

Behind the scenes, an HDInsight cluster exposes a *WebHCat* endpoint. WebHCat is a *Representational State Transfer (REST)*-based API that provides metadata management and remote job submission to the Hadoop cluster. WebHCat is also referred to as *Templeton*. For detailed documentation on Templeton classes and job submissions, refer to the following link:

```
http://docs.hortonworks.com/HDPDocuments/HDP1/HDP-Win-1.1/ds_Templeton/index.html
```

Submitting the wordcount MapReduce Job

The .NET SDK for HDInsight also provides simpler ways to execute your existing MapReduce programs or MapReduce code written in Java. In this section, you will submit and execute the sample wordcount MapReduce job and display the output from the blob storage.

First, let's add a helper function that will wait and display a status when the MapReduce job is in progress. This is important because the MapReduce function calls might not be symmetric and you might see incorrect or intermediate output if you fetch the blob storage when the job execution is in progress. Add the WaitForJobCompletion() method to your Program.cs file with code as shown in Listing 5-8.

Listing 5-8. The WaitForJobCompletion() method

```
private static void WaitForJobCompletion(JobCreationResults jobResults, IJobSubmissionClient client)
        {
            JobDetails jobInProgress = client.GetJob(jobResults.JobId);
            while (jobInProgress.StatusCode != JobStatusCode.Completed &&
                    jobInProgress.StatusCode != JobStatusCode.Failed)
            {
                jobInProgress = client.GetJob(jobInProgress.JobId);
                Thread.Sleep(TimeSpan.FromSeconds(1));
                Console.Write(".");
            }
        }
```

Then add the DoMapReduce() function in your Program.cs file. This function will have the actual code to submit the wordcount job.

The first step is to create the job definition and configure the input and output parameters for the job. This is done using the MapReduceJobCreateParameters class.

```
// Define the MapReduce job
MapReduceJobCreateParameters mrJobDefinition = new MapReduceJobCreateParameters()
        {
                JarFile = "wasb:///example/jars/hadoop-examples.jar",
                ClassName = "wordcount"
        };
mrJobDefinition.Arguments.Add("wasb:///example/data/gutenberg/davinci.txt");
mrJobDefinition.Arguments.Add("wasb:///example/data/WordCountOutput");
```

The next step, as usual, is to grab the correct certificate credentials based on the thumbprint:

```
var store = new X509Store();
store.Open(OpenFlags.ReadOnly);
var cert = store.Certificates.Cast<X509Certificate2>().First(item
   => item.Thumbprint == Constants.thumbprint);
var creds = new JobSubmissionCertificateCredential(Constants.subscriptionId,
   cert, Constants.clusterName);
```

Once the credentials are created, it is time to create a JobSubmissionClient object and call the MapReduce job based on the definition:

```
// Create a hadoop client to connect to HDInsight
var jobClient = JobSubmissionClientFactory.Connect(creds);
```

```
// Run the MapReduce job
JobCreationResults mrJobResults = jobClient.CreateMapReduceJob(mrJobDefinition);
Console.Write("Executing WordCount MapReduce Job.");
// Wait for the job to complete
WaitForJobCompletion(mrJobResults, jobClient);
```

The final step after the job submission is to read and display the stream of output from the blob storage. The following piece of code does that:

```
Stream stream = new MemoryStream();
            CloudStorageAccount storageAccount = CloudStorageAccount.Parse(
                "DefaultEndpointsProtocol=https;AccountName="
                + Constants.storageAccount
                + ";AccountKey="
                + Constants.storageAccountKey);
            CloudBlobClient blobClient = storageAccount.CreateCloudBlobClient();
            CloudBlobContainer blobContainer =
                blobClient.GetContainerReference(Constants.container);
            CloudBlockBlob blockBlob =  blobContainer.GetBlockBlobReference("example/data/
WordCountOutput/part-r-00000");

            blockBlob.DownloadToStream(stream);
            stream.Position = 0;
            StreamReader reader = new StreamReader(stream);
            Console.Write("Done..Word counts are:\n");
            Console.WriteLine(reader.ReadToEnd());
```

The entire DoMapReduce() method should look similar to Listing 5-9.

Listing 5-9. DoMapReduce() method

```
public static void DoMapReduce()
        {
            // Define the MapReduce job
            MapReduceJobCreateParameters mrJobDefinition = new MapReduceJobCreateParameters()
            {
                JarFile = "wasb:///example/jars/hadoop-examples.jar",
                ClassName = "wordcount"
            };
            mrJobDefinition.Arguments.Add("wasb:///example/data/gutenberg/davinci.txt");
            mrJobDefinition.Arguments.Add("wasb:///example/data/WordCountOutput");

            //Get certificate
            var store = new X509Store();
            store.Open(OpenFlags.ReadOnly);
            var cert = store.Certificates.Cast<X509Certificate2>().First(item
                => item.Thumbprint == Constants.thumbprint);
            var creds = new JobSubmissionCertificateCredential(
                Constants.subscriptionId, cert, Constants.clusterName);

            // Create a hadoop client to connect to HDInsight
            var jobClient = JobSubmissionClientFactory.Connect(creds);
```

```
            // Run the MapReduce job
            JobCreationResults mrJobResults = jobClient.CreateMapReduceJob(mrJobDefinition);
            Console.Write("Executing WordCount MapReduce Job.");

            // Wait for the job to complete
            WaitForJobCompletion(mrJobResults, jobClient);

            // Print the MapReduce job output
            Stream stream = new MemoryStream();
            CloudStorageAccount storageAccount =
CloudStorageAccount.Parse("DefaultEndpointsProtocol=https;AccountName=" +
Constants.storageAccount + ";AccountKey=" + Constants.storageAccountKey);
            CloudBlobClient blobClient = storageAccount.CreateCloudBlobClient();
            CloudBlobContainer blobContainer =
blobClient.GetContainerReference(Constants.container);
            CloudBlockBlob blockBlob =
blobContainer.GetBlockBlobReference("example/data/WordCountOutput/part-r-00000");
            blockBlob.DownloadToStream(stream);
            stream.Position = 0;
            StreamReader reader = new StreamReader(stream);
            Console.Write("Done..Word counts are:\n");
            Console.WriteLine(reader.ReadToEnd());
        }
```

Add a call to this method in Program.cs and run the program. You should see the job completing with success, and the words with their counts should be displayed in the console. Thus, the .NET Framework exposes two different ways to submit MapReduce jobs to your HDInsight clusters: you can write your own .NET MapReduce classes, or you can choose to run any of the existing ones bundled in .jar files.

Submitting a Hive Job

As stated earlier, Hive is an abstraction over MapReduce that provides a SQL-like language that is internally broken down to MapReduce jobs. This relieves the programmer of writing the code and developing the MapReduce infrastructure as described in the previous section.

Adding the References

Launch the NuGet *Package Manager Console*, and import the Hive NuGet package by running the following command:

```
install-package Microsoft.Hadoop.Hive
```

This should import the required .dll, along with any dependencies it may have. You will see output similar to the following:

```
PM> install-package Microsoft.Hadoop.Hive
Attempting to resolve dependency 'Newtonsoft.Json (≥ 4.5.11)'.
Installing 'Microsoft.Hadoop.Hive 0.9.4951.25594'.
Successfully installed 'Microsoft.Hadoop.Hive 0.9.4951.25594'.
Adding 'Microsoft.Hadoop.Hive 0.9.4951.25594' to HadoopClient.
```

```
Successfully added 'Microsoft.Hadoop.Hive 0.9.4951.25594' to HadoopClient.
Setting MRLib items CopyToOutputDirectory=true
```

Once the NuGet package has been added, add a reference to the .dll file in your code:

```
using Microsoft.Hadoop.Hive;
```

Once the references are added, you can develop the application code to construct and execute Hive queries against your HDInsight cluster.

Creating the Hive Queries

The Hive .NET API exposes a few key methods to create and run Hive jobs. The steps are pretty similar to creating a MapReduce job submission. Add a new DoHiveOperations() method in your Program.cs file. This method will contain your Hive job submission code.

As with your MapReduce job submission code, the first step is to create your Hive job definition:

```
HiveJobCreateParameters hiveJobDefinition = new HiveJobCreateParameters()
            {
                JobName = "Show tables job",
                StatusFolder = "/TableListFolder",
                Query = "show tables;"
            };
```

Next is the regular piece of code dealing with certificates and credentials to submit and run jobs in the cluster:

```
            var store = new X509Store();
            store.Open(OpenFlags.ReadOnly);
            var cert = store.Certificates.Cast<X509Certificate2>().First(item =>
item.Thumbprint == Constants.thumbprint);
            var creds = new JobSubmissionCertificateCredential(Constants.subscriptionId,
cert, Constants.clusterName);
```

Then create a job submission client object and submit the Hive job based on the definition:

```
var jobClient = JobSubmissionClientFactory.Connect(creds);
JobCreationResults jobResults = jobClient.CreateHiveJob(hiveJobDefinition);
Console.Write("Executing Hive Job.");
// Wait for the job to complete
WaitForJobCompletion(jobResults, jobClient);
```

Finally, you are ready to read the blob storage and display the output:

```
// Print the Hive job output
System.IO.Stream stream = jobClient.GetJobOutput(jobResults.JobId);
System.IO.StreamReader reader = new System.IO.StreamReader(stream);
Console.Write("Done..List of Tables are:\n");
Console.WriteLine(reader.ReadToEnd());
```

Listing 5-10 shows the complete DoHiveOperations() method. Note that it uses the same WaitForJobCompletion() method to wait and display progress while the job execution is in progress.

Listing 5-10. DoHiveOperations() method

```
public static void DoHiveOperations()
        {
            HiveJobCreateParameters hiveJobDefinition = new HiveJobCreateParameters()
            {
                JobName = "Show tables job",
                StatusFolder = "/TableListFolder",
                Query = "show tables;"
            };

            var store = new X509Store();
            store.Open(OpenFlags.ReadOnly);
            var cert = store.Certificates.Cast<X509Certificate2>().First(item
                => item.Thumbprint == Constants.thumbprint);
            var creds = new JobSubmissionCertificateCredential(
                Constants.subscriptionId, cert, Constants.clusterName);
            var jobClient = JobSubmissionClientFactory.Connect(creds);
            JobCreationResults jobResults = jobClient.CreateHiveJob(hiveJobDefinition);
            Console.Write("Executing Hive Job.");
            // Wait for the job to complete
            WaitForJobCompletion(jobResults, jobClient);
            // Print the Hive job output
            System.IO.Stream stream = jobClient.GetJobOutput(jobResults.JobId);
            System.IO.StreamReader reader = new System.IO.StreamReader(stream);
            Console.Write("Done..List of Tables are:\n");
            Console.WriteLine(reader.ReadToEnd());
        }
```

Once this is done, you are ready to submit the Hive job to your cluster.

Running the Hive Job

The final step is to add a call to the DoHiveOperations() method in the Main() function. The Main() method should now look similar to the following:

```
static void Main(string[] args)
        {
            //ListClusters();
            //CreateCluster();
            //DeleteCluster();
            //DoCustomMapReduce();
            //DoMapReduce();
            DoHiveOperations();
            Console.Write("Press any key to exit");
            Console.ReadKey();
        }
```

▨ **Note** You may need to comment out a few of the other function calls to avoid repetitive operations.

Execute the code and you should see output similar to Listing 5-11.

Listing 5-11. Hive job output

```
Executing Hive Job.........Done..List of Tables are:
aaplstockdata
hivesampletable
stock_analysis
stock_analysis1
```

▨ **Note** The `hivesampletable` is the only table that comes built in as a sample. I have other tables created, so your output may be different based on the Hive tables you have.

The .NET APIs provide the .NET developers the flexibility to use their existing skills to automate job submissions in Hadoop. This simple console application can be further enhanced to create a `Windows Form` application and provide a really robust monitoring and job submission interface for your HDInsight clusters.

Monitoring Job Status

The .NET SDK also supports the Hadoop supporting package *Ambari*. Ambari is a framework that provides monitoring and instrumentation options for your cluster. To implement the Ambari APIs, you need to add the NuGet package `Microsoft.Hadoop.WebClient`. You will also need to import the following namespaces in your `Program.cs` file:

```
using Microsoft.Hadoop.WebClient.AmbariClient;
using Microsoft.Hadoop.WebClient.AmbariClient.Contracts;
```

Once the references are added, create a new function called `MonitorCluster()` and add the code snippet as shown in Listing 5-12.

Listing 5-12. MonitorCluster() method

```
public static void MonitorCluster()
        {
              var client = new AmbariClient(Constants.azureClusterUri,
                  Constants.clusterUser, Constants.clusterPassword);
              IList<ClusterInfo> clusterInfos = client.GetClusters();
              ClusterInfo clusterInfo = clusterInfos[0];
              Console.WriteLine("Cluster Href: {0}", clusterInfo.Href);

              Regex clusterNameRegEx = new Regex(@"(\w+)\.*");
              var clusterName = clusterNameRegEx.Match(Constants.azureClusterUri.Authority).Groups[1].
Value;
              HostComponentMetric hostComponentMetric = client.GetHostComponentMetric(
                  clusterName + ".azurehdinsight.net");
```

```
        Console.WriteLine("Cluster Map Reduce Metrics:");
        Console.WriteLine("\tMaps Completed: \t{0}", hostComponentMetric.MapsCompleted);
        Console.WriteLine("\tMaps Failed: \t{0}", hostComponentMetric.MapsFailed);
        Console.WriteLine("\tMaps Killed: \t{0}", hostComponentMetric.MapsKilled);
        Console.WriteLine("\tMaps Launched: \t{0}", hostComponentMetric.MapsLaunched);
        Console.WriteLine("\tMaps Running: \t{0}", hostComponentMetric.MapsRunning);
        Console.WriteLine("\tMaps Waiting: \t{0}", hostComponentMetric.MapsWaiting);
    }
```

When you execute the `MonitorCluster()` method, you should see output similar to the following:

```
Cluster Href: https://democluster.azurehdinsight.net/ambari/api/monitoring/v1/
clusters/democluster.azurehdinsight.net
Cluster Map Redeuce Metrics:
        Maps Completed: 151
        Maps Failed    : 20
        Maps Killed    : 0
        Maps Launched : 171
        Maps Running   : 0
        Maps Waiting   : 10
```

The Ambari APIs can be used as mentioned to display MapReduce metrics for your cluster. The .NET SDK also supports other functionalities, like data serialization using the Open Source Apache project *Avro*. For a complete list of the SDK functionalities, refer to the following site:

```
http://hadoopsdk.codeplex.com/
```

Through the HadoopClient program, we automated MapReduce and Hive job submissions. Bundled together with the cluster-management operations in the previous chapter, the complete Program.cs file along with the using statements should now look similar to Listing 5-13.

Listing 5-13. The complete code listing

```
using System;
using System.Collections.Generic;
using System.Linq;
using System.Text;
using System.Security.Cryptography.X509Certificates;
using Microsoft.WindowsAzure.Management.HDInsight;
using Microsoft.Hadoop.MapReduce;
using Microsoft.Hadoop.Client;
//For Stream IO
using System.IO;
//For Ambari Monitoring Client
using Microsoft.Hadoop.WebClient.AmbariClient;
using Microsoft.Hadoop.WebClient.AmbariClient.Contracts;
//For Regex
using System.Text.RegularExpressions;
//For thread
using System.Threading;
```

```csharp
//For Blob Storage
using Microsoft.WindowsAzure.Storage;
using Microsoft.WindowsAzure.Storage.Blob;

namespace HadoopClient
{
    class Program
    {
        static void Main(string[] args)
        {
            ListClusters();
            CreateCluster();
            DeleteCluster();
            DoCustomMapReduce();
            DoMapReduce();
            DoHiveOperations();
            MonitorCluster();
            Console.Write("Press any key to exit");
            Console.ReadKey();
        }

        //List existing HDI clusters
        public static void ListClusters()
        {
            var store = new X509Store();
            store.Open(OpenFlags.ReadOnly);
            var cert = store.Certificates.Cast<X509Certificate2>().First(item
                => item.Thumbprint == Constants.thumbprint);
            var creds = new HDInsightCertificateCredential(Constants.subscriptionId, cert);
            var client = HDInsightClient.Connect(creds);
            var clusters = client.ListClusters();
            Console.WriteLine("The list of clusters and their details are");
            foreach (var item in clusters)
            {
                Console.WriteLine("Cluster: {0}, Nodes: {1}, State: {2}, Version: {3}",
                    item.Name, item.ClusterSizeInNodes, item.State, item.Version);
            }
        }
        //Create a new HDI cluster
        public static void CreateCluster()
        {
            var store = new X509Store();
            store.Open(OpenFlags.ReadOnly);
            var cert = store.Certificates.Cast<X509Certificate2>().First(item
                => item.Thumbprint == Constants.thumbprint);
            var creds = new HDInsightCertificateCredential(Constants.subscriptionId, cert);
            var client = HDInsightClient.Connect(creds);
```

```
    //Cluster information
  var clusterInfo = new ClusterCreateParameters()
   {
       Name = "AutomatedHDICluster",
       Location = "North Europe",
       DefaultStorageAccountName = Constants.storageAccount,
       DefaultStorageAccountKey = Constants.storageAccountKey,
       DefaultStorageContainer = Constants.container,
       UserName = Constants.clusterUser,
       Password = Constants.clusterPassword,
       ClusterSizeInNodes = 2,
       Version="2.1"
   };
  Console.Write("Creating cluster...");
  var clusterDetails = client.CreateCluster(clusterInfo);
  Console.Write("Done\n");
  ListClusters();
}

//Delete an existing HDI cluster
public static void DeleteCluster()
{
    var store = new X509Store();
    store.Open(OpenFlags.ReadOnly);
    var cert = store.Certificates.Cast<X509Certificate2>().First(item
       => item.Thumbprint == Constants.thumbprint);
    var creds = new HDInsightCertificateCredential(Constants.subscriptionId, cert);
    var client = HDInsightClient.Connect(creds);
    Console.Write("Deleting cluster...");
    client.DeleteCluster("AutomatedHDICluster");
    Console.Write("Done\n");
    ListClusters();
}

//Run Custom Map Reduce
public static void DoCustomMapReduce()
{
    Console.WriteLine("Starting MapReduce job. Log in remotely to your Name Node " +
       "and check progress from JobTracker portal with the returned JobID…");
    IHadoop hadoop = Hadoop.Connect(Constants.azureClusterUri, Constants.clusterUser,
       Constants.hadoopUser, Constants.clusterPassword, Constants.storageAccount,
       Constants.storageAccountKey, Constants.container, true);
    var output = hadoop.MapReduceJob.ExecuteJob<SquareRootJob>();
}
//Run Sample Map Reduce Job
public static void DoMapReduce()
{
    // Define the MapReduce job
    MapReduceJobCreateParameters mrJobDefinition = new MapReduceJobCreateParameters()
    {
        JarFile = "wasb:///example/jars/hadoop-examples.jar",
        ClassName = "wordcount"
    };
```

```
        mrJobDefinition.Arguments.Add("wasb:///example/data/gutenberg/davinci.txt");
        mrJobDefinition.Arguments.Add("wasb:///example/data/WordCountOutput");

        //Get certificate
        var store = new X509Store();
        store.Open(OpenFlags.ReadOnly);
        var cert = store.Certificates.Cast<X509Certificate2>().First(item
            => item.Thumbprint == Constants.thumbprint);
        var creds = new JobSubmissionCertificateCredential(Constants.subscriptionId,
            cert, Constants.clusterName);

        // Create a hadoop client to connect to HDInsight
        var jobClient = JobSubmissionClientFactory.Connect(creds);

        // Run the MapReduce job
        JobCreationResults mrJobResults = jobClient.CreateMapReduceJob(mrJobDefinition);
        Console.Write("Executing WordCount MapReduce Job.");

        // Wait for the job to complete
        Wai

        // Print the MapReduce job output
        Stream stream = new MemoryStream();
        CloudStorageAccount storageAccount =
CloudStorageAccount.Parse("DefaultEndpointsProtocol=https;AccountName=" +
Constants.storageAccount + ";AccountKey=" + Constants.storageAccountKey);
        CloudBlobClient blobClient = storageAccount.CreateCloudBlobClient();
        CloudBlobContainer blobContainer =
blobClient.GetContainerReference(Constants.container);
        CloudBlockBlob blockBlob =
blobContainer.GetBlockBlobReference("example/data/WordCountOutput/part-r-00000");
        blockBlob.DownloadToStream(stream);
        stream.Position = 0;
        StreamReader reader = new StreamReader(stream);
        Console.Write("Done..Word counts are:\n");
        Console.WriteLine(reader.ReadToEnd());
    }

    //Run Hive Job
    public static void DoHiveOperations()
    {
        HiveJobCreateParameters hiveJobDefinition = new HiveJobCreateParameters()
        {
            JobName = "Show tables job",
            StatusFolder = "/TableListFolder",
            Query = "show tables;"
        };

        var store = new X509Store();
        store.Open(OpenFlags.ReadOnly);
        var cert = store.Certificates.Cast<X509Certificate2>().First(item
            => item.Thumbprint == Constants.thumbprint);
```

```csharp
            var creds = new JobSubmissionCertificateCredential(Constants.subscriptionId,
                cert, Constants.clusterName);
            var jobClient = JobSubmissionClientFactory.Connect(creds);
            JobCreationResults jobResults = jobClient.CreateHiveJob(hiveJobDefinition);
            Console.Write("Executing Hive Job.");
            // Wait for the job to complete
            WaitForJobCompletion(jobResults, jobClient);
            // Print the Hive job output
            System.IO.Stream stream = jobClient.GetJobOutput(jobResults.JobId);
            System.IO.StreamReader reader = new System.IO.StreamReader(stream);
            Console.Write("Done..List of Tables are:\n");
            Console.WriteLine(reader.ReadToEnd());
        }
        //Monitor cluster Map Reduce statistics
        public static void MonitorCluster()
        {
            var client = new AmbariClient(Constants.azureClusterUri,
                Constants.clusterUser, Constants.clusterPassword);
            IList<ClusterInfo> clusterInfos = client.GetClusters();
            ClusterInfo clusterInfo = clusterInfos[0];
            Console.WriteLine("Cluster Href: {0}", clusterInfo.Href);

            Regex clusterNameRegEx = new Regex(@"(\w+)\.*");
            var clusterName =
clusterNameRegEx.Match(Constants.azureClusterUri.Authority).Groups[1].Value;
            HostComponentMetric hostComponentMetric = client.GetHostComponentMetric(
                clusterName + ".azurehdinsight.net");
            Console.WriteLine("Cluster Map Reduce Metrics:");
            Console.WriteLine("\tMaps Completed: \t{0}", hostComponentMetric.MapsCompleted);
            Console.WriteLine("\tMaps Failed: \t{0}", hostComponentMetric.MapsFailed);
            Console.WriteLine("\tMaps Killed: \t{0}", hostComponentMetric.MapsKilled);
            Console.WriteLine("\tMaps Launched: \t{0}", hostComponentMetric.MapsLaunched);
            Console.WriteLine("\tMaps Running: \t{0}", hostComponentMetric.MapsRunning);
            Console.WriteLine("\tMaps Waiting: \t{0}", hostComponentMetric.MapsWaiting);
        }

        ///Helper Function to Wait while job executes
        private static void WaitForJobCompletion(JobCreationResults jobResults,
            IJobSubmissionClient client)
        {
            JobDetails jobInProgress = client.GetJob(jobResults.JobId);
            while (jobInProgress.StatusCode != JobStatusCode.Completed &&
                    jobInProgress.StatusCode != JobStatusCode.Failed)
            {
                jobInProgress = client.GetJob(jobInProgress.JobId);
                Thread.Sleep(TimeSpan.FromSeconds(1));
                Console.Write(".");
            }
        }

    }
}
```

■ **Note**   Do not forget the supporting MapReduce classes `SquareRootMapper`, `SquareRootReducer`, `SquareRootJob`, and `Constants`.

Using PowerShell

Apart from the .NET Framework, HDInsight also supports PowerShell cmdlets for job submissions. As of this writing, the Azure HDInsight cmdlets are available as a separate download from the Microsoft download center. In the future, it will be a part of *Windows Azure PowerShell version 0.7.2* and there will be no separate download. *Windows Azure HDInsight PowerShell* can be downloaded from:

```
http://www.windowsazure.com/en-us/documentation/articles/hdinsight-install-configure-powershell/
```

Writing Script

For better code management and readability, let's define a few PowerShell variables to store the path of the `.dll` files you will refer to throughout the script:

```
$subscription = "Your_Subscription_Name"
$cluster = "democluster"
$storageAccountName = "democluster"
$Container = "democlustercontainer"
$storageAccountKey = Get-AzureStorageKey $storageAccountName | %{ $_.Primary }
$storageContext = New-AzureStorageContext –StorageAccountName $storageAccountName
   -StorageAccountKey $storageAccountKey
$inputPath = "wasb:///example/data/gutenberg/davinci.txt"
$outputPath = "wasb:///example/data/WordCountOutputPS"
$jarFile = "wasb:///example/jars/hadoop-examples.jar"
$class = "wordcount"
$secpasswd = ConvertTo-SecureString "Your_Password" -AsPlainText -Force
$myCreds = New-Object System.Management.Automation.PSCredential ("admin", $secpasswd)
```

The sequence of operations needed to move you toward a job submission through PowerShell is pretty much the same as in the .NET client:

- Creating the job definition

- Submitting the job

- Waiting for the job to complete

- Reading and displaying the output

The following piece of PowerShell script does that in sequence:

```
# Define the word count MapReduce job
$mapReduceJobDefinition = New-AzureHDInsightMapReduceJobDefinition -JarFile $jarFile -ClassName
$class  -Arguments $inputPath, $outputPath

# Submit the MapReduce job
Select-AzureSubscription $subscription
```

```
$wordCountJob = Start-AzureHDInsightJob -Cluster $cluster -JobDefinition $mapReduceJobDefinition
-Credential $myCreds

# Wait for the job to complete
Wait-AzureHDInsightJob -Job $wordCountJob -WaitTimeoutInSeconds 3600 -Credential $myCreds

# Get the job standard error output
Get-AzureHDInsightJobOutput -Cluster $cluster -JobId $wordCountJob.JobId -StandardError
-Subscription $subscription

# Get the blob content
Get-AzureStorageBlobContent -Container $Container -Blob example/data/WordCountOutputPS /part-r-00000
-Context $storageContext -Force

# List the content of the output file
cat ./example/data/WordCountOutputPS/part-r-00000 | findstr "human"
```

▪ **Note** Because the output would be a huge number of words and their counts, we would display only the words that have the string *human* in it.

As you continue to develop your script-based framework for job submissions, it becomes increasingly difficult to manage it without a standard editor. The Windows Azure PowerShell kit provides you with a development environment called *Windows PowerShell ISE*, which makes it easy to write, execute, and debug PowerShell scripts. Figure 5-6 shows you a glimpse of PowerShell ISE. It has built-in IntelliSense and autocomplete features for your variable or method names that comes into play as you type in your code. It also implements a standard coloring mechanism that helps you visually distinguish between the different PowerShell object types.

```
SubmitJob.ps1* X   Invoke-Hive.ps1   Untitled1.ps1*
  1    $subscription = "Your_subscription_Name"
  2    $cluster = "democluster"
  3    $storageAccountName = "democluster"
  4    $Container = "democlustercontainer"
  5    $storageAccountKey = Get-AzureStorageKey $storageAccountName | %{ $_.Primary }
  6    $storageContext = New-AzureStorageContext -StorageAccountName $storageAccountName -StorageAcc
  7    $inputPath = "wasb:///example/data/gutenberg/davinci.txt"
  8    $outputPath = "wasb:///example/data/WordCountOutputPS"
  9    $jarFile = "wasb:///example/jars/hadoop-examples.jar"
 10    $class = "wordcount"
 11    $passwd = ConvertTo-SecureString "XXXXXXXXXXXXXXXX" -AsPlainText -Force
 12    $myCreds = New-Object System.Management.Automation.PSCredential ("admin", $secpasswd)
 13
 14    # Define the word count MapReduce job
 15    $mapReduceJobDefinition = New-AzureHDInsightMapReduceJobDefinition -JarFile $jarFile -ClassNam
 16
 17    # Submit the MapReduce job
 18    Select-AzureSubscription $subscription
 19    $wordCountJob = Start-AzureHDInsightJob -Cluster $cluster -JobDefinition $mapReduceJobDefiniti
 20
 21    # Wait for the job to complete
 22    Wait-AzureHDInsightJob -Job $wordCountJob -WaitTimeoutInSeconds 3600 -Credential $myCreds
 23
 24    # Get the job standard error output
 25    Get-AzureHDInsightJobOutput -Cluster $cluster -JobId $wordCountJob.JobId -StandardError -Subsc
 26
 27    # Get the blob content
 28    Get-AzureStorageBlobContent -Container $Container -Blob example/data/WordCountOutputPS/part-r-
 29
 30    # List the content of the output file
 31    cat ./example/data/WordCountOutputPS/part-r-00000 | findstr "human"
```

```
PS D:\HadoopClient>
```

Figure 5-6. *Windows PowerShell ISE*

The entire script can be saved as a PowerShell script file (.ps1) for later execution. Listing 5-14 shows the complete script.

Listing 5-14. PowerShell job submission script

```
$subscription = "Your_Subscription_Name"
$cluster = "democluster"
$storageAccountName = "democluster"
$Container = "democlustercontainer"
$storageAccountKey = Get-AzureStorageKey $storageAccountName | %{ $_.Primary }
$storageContext = New-AzureStorageContext -StorageAccountName $storageAccountName
   -StorageAccountKey $storageAccountKey
```

```
$inputPath = "wasb:///example/data/gutenberg/davinci.txt"
$outputPath = "wasb:///example/data/WordCountOutput"
$jarFile = "wasb:///example/jars/hadoop-examples.jar"
$class = "wordcount"
$passwd = ConvertTo-SecureString "Your_Password" -AsPlainText -Force
$myCreds = New-Object System.Management.Automation.PSCredential ("admin", $secpasswd)

# Define the word count MapReduce job
$mapReduceJobDefinition = New-AzureHDInsightMapReduceJobDefinition -JarFile $jarFile
    -ClassName $class  -Arguments $inputPath, $outputPath

# Submit the MapReduce job
Select-AzureSubscription $subscription
$wordCountJob = Start-AzureHDInsightJob -Cluster $cluster -JobDefinition
    $mapReduceJobDefinition -Credential $myCreds

# Wait for the job to complete
Wait-AzureHDInsightJob -Job $wordCountJob -WaitTimeoutInSeconds 3600 -Credential $myCreds

# Get the job standard error output
Get-AzureHDInsightJobOutput -Cluster $cluster -JobId $wordCountJob.JobId -StandardError
    -Subscription $subscription

# Get the blob content
Get-AzureStorageBlobContent -Container $Container -Blob example/data/WordCountOutputPS/part-r-00000
-Context $storageContext -Force

# List the content of the output file
cat ./example/data/WordCountOutputPS/part-r-00000 | findstr "human"
```

Executing The Job

You can execute the script directly from PowerShell ISE or use the Windows Azure PowerShell command prompt.
Save the script file as SubmitJob.ps1 in a location of your choice, and execute it from the PowerShell prompt. You
should see an output similar to the following once the script completes successfully:

```
PS C:\> C:\SubmitJob.ps1
StatusDirectory : 0fac8406-891d-41ff-af74-eaac21386fd3
ExitCode        : 0
Name            : wordcount
Query           :
State           : Completed
SubmissionTime  : 12/9/2013 7:47:05 PM
Cluster         : democluster
PercentComplete : map 100% reduce 100%
JobId           : job_201311240635_0192

13/12/09 19:47:19 INFO input.FileInputFormat: Total input paths to process : 1
13/12/09 19:47:19 WARN snappy.LoadSnappy: Snappy native library is available
13/12/09 19:47:19 INFO util.NativeCodeLoader: Loaded the native-hadoop library
```

```
13/12/09 19:47:19 INFO snappy.LoadSnappy: Snappy native library loaded
13/12/09 19:47:19 INFO mapred.JobClient:  Running job: job_201311240635_0193
13/12/09 19:47:20 INFO mapred.JobClient:  map 0% reduce 0%
13/12/09 19:47:29 INFO mapred.JobClient:  map 100% reduce 0%
13/12/09 19:47:37 INFO mapred.JobClient:  map 100% reduce 33%
13/12/09 19:47:39 INFO mapred.JobClient:  map 100% reduce 100%
13/12/09 19:47:41 INFO mapred.JobClient:  Job complete: job_201311240635_0193
13/12/09 19:47:42 INFO mapred.JobClient:  Counters: 30
13/12/09 19:47:42 INFO mapred.JobClient:  Job Counters
13/12/09 19:47:42 INFO mapred.JobClient:  Launched reduce tasks=1
13/12/09 19:47:42 INFO mapred.JobClient:  SLOTS_MILLIS_MAPS=8500
13/12/09 19:47:42 INFO mapred.JobClient:
 Total time spent by all reduces waiting after reserving slots (ms)=0
13/12/09 19:47:42 INFO mapred.JobClient:
Total time spent by all maps waiting after reserving slots (ms)=0
13/12/09 19:47:42 INFO mapred.JobClient:    Rack-local map tasks=1
13/12/09 19:47:42 INFO mapred.JobClient:    Launched map tasks=1
13/12/09 19:47:42 INFO mapred.JobClient:    SLOTS_MILLIS_REDUCES=10640
13/12/09 19:47:42 INFO mapred.JobClient:    File Output Format Counters
13/12/09 19:47:42 INFO mapred.JobClient:    Bytes Written=337623
13/12/09 19:47:42 INFO mapred.JobClient:    FileSystemCounters
13/12/09 19:47:42 INFO mapred.JobClient:    WASB_BYTES_READ=1395666
13/12/09 19:47:42 INFO mapred.JobClient:    FILE_BYTES_READ=466915
13/12/09 19:47:42 INFO mapred.JobClient:    HDFS_BYTES_READ=161
13/12/09 19:47:42 INFO mapred.JobClient:    FILE_BYTES_WRITTEN=1053887
13/12/09 19:47:42 INFO mapred.JobClient:    WASB_BYTES_WRITTEN=337623
13/12/09 19:47:42 INFO mapred.JobClient:    File Input Format Counters
13/12/09 19:47:42 INFO mapred.JobClient:    Bytes Read=1395667
13/12/09 19:47:42 INFO mapred.JobClient:    Map-Reduce Framework
13/12/09 19:47:42 INFO mapred.JobClient:    Map output materialized bytes=466761
13/12/09 19:47:42 INFO mapred.JobClient:    Map input records=32118
13/12/09 19:47:42 INFO mapred.JobClient:    Reduce shuffle bytes=466761
13/12/09 19:47:42 INFO mapred.JobClient:    Spilled Records=65912
13/12/09 19:47:42 INFO mapred.JobClient:    Map output bytes=2387798
13/12/09 19:47:42 INFO mapred.JobClient:    Total committed heap usage (bytes)=1029046272
13/12/09 19:47:42 INFO mapred.JobClient:    CPU time spent (ms)=7547
13/12/09 19:47:42 INFO mapred.JobClient:    Combine input records=251357
13/12/09 19:47:42 INFO mapred.JobClient:    SPLIT_RAW_BYTES=161
13/12/09 19:47:42 INFO mapred.JobClient:    Reduce input records=32956
13/12/09 19:47:42 INFO mapred.JobClient:    Reduce input groups=32956
13/12/09 19:47:42 INFO mapred.JobClient:    Combine output records=32956
13/12/09 19:47:42 INFO mapred.JobClient:    Physical memory (bytes) snapshot=495923200
13/12/09 19:47:42 INFO mapred.JobClient:    Reduce output records=32956
13/12/09 19:47:42 INFO mapred.JobClient:    Virtual memory (bytes) snapshot=1430675456
13/12/09 19:47:42 INFO mapred.JobClient:    Map output records=251357

ICloudBlob   : Microsoft.WindowsAzure.Storage.Blob.CloudBlockBlob
BlobType     : BlockBlob
Length       : 337623
ContentType  : application/octet-stream
LastModified : 12/9/2013 7:47:39 PM +00:00
```

```
SnapshotTime :
Context       : Microsoft.WindowsAzure.Commands.Storage.Model.ResourceModel.AzureStorageContext
Name          : example/data/WordCountOutputPS/part-r-00000

human       57
humana,      1
humane,      1
humani       2
humanist).   1
humanists    1
humanorum    1
inhuman      1
l'humano     1
```

Depending on your computer's security policies, you may get an exception, as shown next, when you try to run PowerShell scripts that use .dll files that are compiled and signed externally:

```
PS C:\> .\SubmitJob.ps1
.\SubmitJob.ps1 : File C:\SubmitJob.ps1 cannot be loaded because running scripts is disabled on this
system. For more
information, see about_Execution_Policies at http://go.microsoft.com/fwlink/?LinkID=135170.
At line:1 char:1
+ .\SubmitJob.ps1
```

If you encounter such a problem, you need to explicitly set the PowerShell execution policy using the following command:

```
Set-ExecutionPolicy RemoteSigned
```

While setting the execution policy, accept any warnings you might get in the PowerShell console. It is also possible to submit Hive jobs using PowerShell, much like the .NET SDK. Carl Nolan has a great blog that covers Hive job submission through PowerShell:

```
http://blogs.msdn.com/b/carlnol/archive/2013/06/18/managing-hive-job-submissions-with-powershell.
aspx
```

Using MRRunner

To submit MapReduce jobs, HDInsight distribution offers a command-line utility called *MRRunner*, which could be utilized as well apart from the .NET SDK and the HDInsight PowerShell cmdlets. Again, to support the MRRunner utility, you should have an assembly (a .NET .dll) that defines at least one implementation of HadoopJob<>.

If the .dll contains only one implementation of HadoopJob<>, (like our HadoopClient.dll does), you can run the job with the following:

```
MRRunner -dll MyDll
```

If the .dll contains multiple implementations of HadoopJob<>, you need to indicate the one you wish to run:

```
MRRunner -dll MyDll -class MyClass
```

To supply additional configuration options to your job, you need to pass them as trailing arguments on the command line, after a double-hyphen:

```
MRRunner -dll MyDll -class MyClass -- extraArg1 extraArg2
```

These additional arguments are provided to your job via a context object that is available to all methods on `HadoopJob<>`.

When you develop a project using the .NET SDK, the MRRunner utility will be automatically deployed in a folder called `MRLib` in your project directory, as illustrated in Figure 5-7. It is basically a Windows executable (.exe) file.

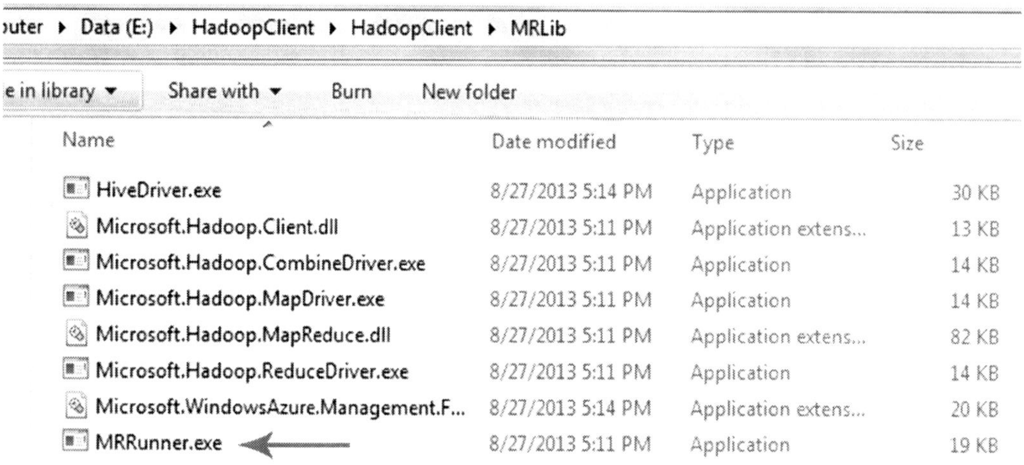
›uter ▸ Data (E:) ▸ HadoopClient ▸ HadoopClient ▸ MRLib
e in library ▾
Share with ▾
Burn
New folder
Name
Date modified
Type
Size
HiveDriver.exe 8/27/2013 5:14 PM Application 30 KB
Microsoft.Hadoop.Client.dll 8/27/2013 5:11 PM Application extens... 13 KB
Microsoft.Hadoop.CombineDriver.exe 8/27/2013 5:11 PM Application 14 KB
Microsoft.Hadoop.MapDriver.exe 8/27/2013 5:11 PM Application 14 KB
Microsoft.Hadoop.MapReduce.dll 8/27/2013 5:11 PM Application extens... 82 KB
Microsoft.Hadoop.ReduceDriver.exe 8/27/2013 5:11 PM Application 14 KB
Microsoft.WindowsAzure.Management.F... 8/27/2013 5:14 PM Application extens... 20 KB
MRRunner.exe 8/27/2013 5:11 PM Application 19 KB

Figure 5-7. *MRRunner.exe utility*

You can launch a command prompt and run the `MRRunner.exe` with appropriate arguments. Specify the `HadoopClient.dll` from the project's `bin\debug` folder as in the following example:

```
E:\HadoopClient\HadoopClient\MRLib>MRRunner -dll "E:\HadoopClient\HadoopClient\bin\Debug\
HadoopClient.dll"
```

■ **Note** In case you are using a release build for your project, you will find the `HadoopClient.dll` file in your project's `bin\release` folder. You also need to change the Project output type to Class Library to generate the `HadoopClient.dll` from the Project ➤ Properties menu.

On successful completion of the job, you will see output similar to Listing 5-15.

Listing 5-15. MRRunner output

```
Output folder exists.. deleting.

File dependencies to include with job:
[Auto-detected] C :\windows\Microsoft.Net\assembly\GAC_MSIL\PresentationFramework\
   v4.0_4.0.0.0__31bf3856ad364e35\PresentationFramework.dll
[Auto-detected] C:\windows\Microsoft.Net\assembly\GAC_32\PresentationCore\
```

```
    v4.0_4.0.0.0__31bf3856ad364e35\PresentationCore.dll
[Auto -detected] C:\windows\Microsoft.Net\assembly\GAC_MSIL\UIAutomationProvider\
    v4.0_4.0.0.0__31bf3856ad364e35\UIAutomationProvider.dll
[Auto-detected] C:\windows\Microsoft.Net\assembly\GAC_MSIL\UIAutomationTypes\
    v4.0_4.0.0.0__31bf3856ad364e35\UIAutomationTypes.dll
[Auto-detected] C:\windows\Microsoft.Net\assembly\GAC_MSIL\PresentationFramework.Aero2\
    v4.0_4.0.0.0__31bf3856ad364e35\PresentationFramework.Aero2.dll
[Auto-detected] C:\windows\Microsoft.Net\assembly\GAC_MSIL\PresentationFramework-SystemXml\
    v4.0_4.0.0.0__b77a5c561934e089\PresentationFramework-SystemXml.dll
[Auto-detected] C:\windows\Microsoft.Net\assembly\GAC_MSIL\PresentationFramework-SystemCore\
    v4.0_4.0.0.0__b77a5c561934e089\PresentationFramework-SystemCore.dll
[Auto-detected] C:\windows\Microsoft.Net\assembly\GAC_MSIL\PresentationFramework-SystemData\
    v4.0_4.0.0.0__b77a5c561934e089\PresentationFramework-SystemData.dll
[Auto-detected] D:\HadoopClient\HadoopClient\bin\Release\microsoft.hadoop.client.dll
[Auto-detected] D:\HadoopClient\HadoopClient\bin\Release\microsoft.hadoop.mapreduce.dll
[Auto-detected] D:\HadoopClient\HadoopClient\bin\Release\microsoft.hadoop.webclient.dll
[Auto-detected] D:\HadoopClient\HadoopClient\bin\Release\Newtonsoft.Json.dll
[Auto-detected] D:\HadoopClient\HadoopClient\bin\Release\HadoopClient.dll
[Auto-detected] C:\windows\Microsoft.Net\assembly\GAC_MSIL\PresentationFramework-SystemXmlLinq\
    v4.0_4.0.0.0__b77a5c561934e089\PresentationFramework-SystemXmlLinq.dll
[Auto-detected] C:\windows\Microsoft.Net\assembly\GAC_MSIL\UIAutomationClient\
    v4.0_4.0.0.0__31bf3856ad364e35\UIAutomationClient.dll
[Auto-detected] C:\windows\Microsoft.Net\assembly\GAC_MSIL\PresentationUI\
    v4.0_4.0.0.0__31bf3856ad364e35\PresentationUI.dll
[Auto-detected] C:\windows\Microsoft.Net\assembly\GAC_MSIL\ReachFramework\
    v4.0_4.0.0.0__31bf3856ad364e35\ReachFramework.dll

Job job_201309210954_0193 completed.
```

The MRRunner command can be put in a Windows batch file (.bat) or a command file (.cmd) and scheduled in Windows Task Scheduler to execute it on a periodic basis. Of course, there are plenty of other ways as well to automate MRRunner operations.

Summary

One of the major benefits of using the Azure HDInsight service is the elasticity it provides in terms of spinning up clusters and running jobs exactly when they are required. The basic idea behind this is to avoid preserving idle clusters just for storage. In HDInsight, the ultimate goal will be to present a script or a program that demonstrates how you can provide a DLL and have the script bring a cluster online, run your job, and then remove the cluster, while allowing you to specify the cluster name and the number of hosts needed to run the job. There are various ways you can provision a new cluster, with the simplest of them being the Management portal. and it's easy-to-use, intuitive graphical user interface. But as requirements become more and more complex and unpredictable along with project budget limitations, automating and parameterizing cluster provisioning and job submissions become a necessity. You can also provision cluster and configure it to connect to more than one Azure Blob storage or custom Hive and Oozie metastores. This advanced feature allows you to separate lifetime of your data and metadata from the lifetime of the cluster. There is a great sample script to provision an HDInsight cluster using custom configuration available at:

```
http://www.windowsazure.com/en-us/documentation/articles/hdinsight-provision-clusters/
```

CHAPTER 6

■ ■ ■

Exploring the HDInsight Name Node

The HDInsight name node is just another virtual machine provisioned in Windows Azure. Theoretically, this is the equivalent of the traditional Apache Hadoop name node or the head node, which is the heart and soul of your Hadoop cluster. I would like to re-iterate what I pointed out in Chapter 1: the name node is the single point of failure in a Hadoop cluster. Most important of all, the name node contains the metadata of the entire cluster storage blocks and maintains co-ordination among the data nodes, so understandably it could bring down the entire cluster.

■ **Note**   There is a *Secondary Name Node* service (ideally run on a dedicated physical server) that keeps track of the changed HDFS blocks in the name node and periodically backs up the name node. In addition, you can fail over to the secondary name in the unlikely event of a name-node failure, but that failover is a manual process.

The HDInsight Service brings a significant change from the traditional approach taken in Apache Hadoop. It does so by isolating the storage to a Windows Azure Storage Blob instead of to the traditional Hadoop Distributed File System (HDFS) that is local to the data nodes.

In the Windows Azure HDInsight service, the storage is separated from the cluster itself by default; the default Hadoop file system is pointed to Azure blob storage rather than traditional HDFS in HDInsight distribution. If you recall, we discussed the advantages of using Windows Azure Storage Blob (WASB) earlier in Chapter 2. This reduces the cluster's dependency on the name node to some extent; still, the HDInsight name node continues to be an integral part of your cluster. You could start a remote desktop session to log on to the name node and get access to the traditional Apache Hadoop web portals and dashboards. This also gives you access to the Hadoop command prompt and the various service logs, and it is the old-fashioned way to administer your cluster.

It continues to be a favorite for a lot of users who still prefer the command-prompt way of doing things in today's world of rich and intuitive user interfaces for almost everything. I often find myself in this category too because I believe command-line interfaces are the bare minimum and they give you the raw power of your modules by getting rid of any abstractions in between. It is also a good practice to operate your cluster using the command shell to test and benchmark performance because it does not have any additional overhead. This chapter focuses on some of the basic command-line utilities to operate your Hadoop cluster and the unique features that are implemented in the HDInsight offering.

Accessing the HDInsight Name Node

You have to enable remote connectivity to your name node from the Azure Management portal. By default, remote login is turned off. You can enable it from your cluster's configuration screen as shown in Figure 6-1.

Figure 6-1. *Enabling Remote Desktop to access the cluster name node*

Create the user to be granted remote desktop access to the name node in the Configure Remote Desktop screen, as shown in Figure 6-2. Be sure to supply a password. You also have to choose an expiration date for this user account. For security reasons, you will need to reconfigure your remote desktop user every seven days. The expiration date that needs to be set will not accept a date greater than a week into the future.

CONFIGURE HDINSIGHT

Configure Remote Desktop

USER NAME

hadoopuser

PASSWORD   CONFIRM PASSWORD

••••••••••••••   ••••••••••••••

EXPIRES ON ❓

2013-12-16

Figure 6-2. *Configuring a Remote Desktop user*

Within a minute or two, Remote Desktop will be enabled for your cluster. You will then see the Connect option as shown in Figure 6-3.

Figure 6-3. *Remote Desktop enabled*

Click on the Connect link. Open the Remote Desktop file `democluster.azurehdinsight.net.rdp`. Accept the couple of security prompts you might get. (Choose not to prompt again.) You will then get a screen where you need to provide the credentials to connect to the name node. Provide the username and password you just created while enabling Remote Desktop for your cluster, as shown in Figure 6-4.

Figure 6-4. *Log on to the name node*

Once valid credentials are provided, you are presented with the desktop of your cluster's name node. HDInsight distribution creates three shortcuts for you, and you will see them on the name node's desktop as shown in Figure 6-5. The shortcuts are

- **Hadoop Command Line**: Invokes the command line, which is the traditional Windows command prompt launched from the `c:\apps\dist\hadoop-1.2.0.1.3.1.0-06\` directory. This is the base for command-line executions of the Hadoop commands, as well as for commands relating to Hive, Pig, Sqoop, and several other supporting projects.

- **Hadoop MapReduce Status**: This is a Java-based web application that comes with the Apache Hadoop distribution. The MapReduce status portal displays the MapReduce configurations based on the config file `mapred-site.xml`. It also shows a history of all the map and reduce task executions in the cluster based on the job id. You can drill down to individual jobs and their tasks to examine a MapReduce job execution.

- **Hadoop Name Node Status**: This is also a Java-based web portal prebuilt in Apache Hadoop. The NameNode status portal displays the file system health as well as the cluster health in terms of the number of live nodes, dead nodes, and decommissioning nodes. You can also navigate through the HDFS and load chunks of data from job output files for display.

Figure 6-5. *The Name Node desktop*

You'll use the command line a lot, so let's look at that next in the sections to follow.

Hadoop Command Line

Traditional Linux-based Apache Hadoop uses shell scripting to implement the commands. Essentially, most of the commands are .sh files that need to be invoked from the command prompt. Hadoop on Windows relies on command files (.cmd) and PowerShell scripts (.ps1) to simulate the command-line shell. HDInsight has unique capabilities to talk to WASB; hence, you can operate natively with your Azure storage account containers in the cloud.

To access the Hadoop command prompt, double-click the shortcut Hadoop Command Line on your name node's desktop. (See Figure 6-6.)

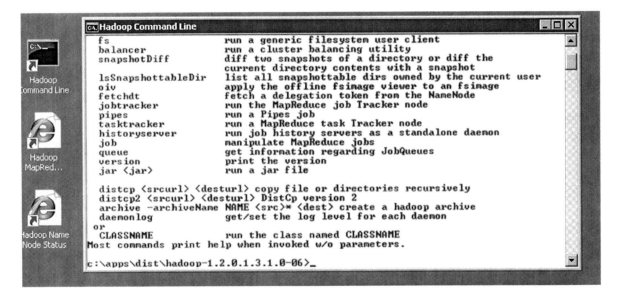

Figure 6-6. *The Hadoop command line*

This will look very familiar to traditional Hadoop users because this is exactly what you find in the Apache Open Source project. Again, the point to be noted here is HDInsight is built on top of core Hadoop, so it supports all the interfaces available with core Hadoop, including the command prompt.

For example, you can run the standard ls command to list the directory and file structure of the current directory. The command in Listing 6-1 lists the files and folders you have in the root of your container.

Listing 6-1. The HDFS directory structure

```
hadoop dfs –ls /
```

This command lists the files and folders in the root of your storage account container in Azure, as shown in Figure 6-7.

```
drwxr-xr-x   - debarchan       supergroup          0 2013-11-24 06:55 /debarchan
drwxr-xr-x   - hdpinternaluser supergroup          0 2013-11-24 06:36 /example
drwxr-xr-x   - hadoopuser      supergroup          0 2013-12-06 15:11 /hadoopuse
r
drwxr-xr-x   - debarchan       supergroup          0 2013-11-24 06:54 /hdp
drwxr-xr-x   - hdp             supergroup          0 2013-11-24 06:36 /hive
drwxr-xr-x   - hdp             supergroup          0 2013-11-24 06:35 /mapred
drwxr-xr-x   - admin           supergroup          0 2013-12-09 13:59 /output1
drw-r--r--   - hdp             supergroup          0 2013-11-24 07:05 /templeton
-hadoop
drwxr-xr-x   - SYSTEM          supergroup          0 2013-11-24 06:35 /user
```

Figure 6-7. *The* ls *command output*

You can run the word-count MapReduce job through the command prompt on the source file provided in the /example/data/gutenburg directory in your WASB to generate the output file much like you did from the .NET and PowerShell code in Chapter 5. The command to invoke the MapReduce job is provided in Listing 6-2.

Listing 6-2. Running the word-count MapReduce job from the Hadoop command line

```
hadoop jar hadoop-examples.jar wordcount /example/data/gutenberg/davinci.txt
/example/data/commandlineoutput
```

This launches the MapReduce job on the input file, and you should see an output similar to Listing 6-3.

Listing 6-3. MapReduce command-line output

```
13/12/09 22:33:42 INFO input.FileInputFormat: Total input paths to process : 1
13/12/09 22:33:42 WARN snappy.LoadSnappy: Snappy native library is available
13/12/09 22:33:42 INFO util.NativeCodeLoader: Loaded the native-hadoop library
13/12/09 22:33:42 INFO snappy.LoadSnappy:    Snappy native library loaded
13/12/09 22:33:43 INFO mapred.JobClient:     Running job: job_201311240635_0196
13/12/09 22:33:44 INFO mapred.JobClient:     map 0% reduce 0%
13/12/09 22:33:55 INFO mapred.JobClient:     map 100% reduce 0%
13/12/09 22:34:03 INFO mapred.JobClient:     map 100% reduce 33%
13/12/09 22:34:05 INFO mapred.JobClient:     map 100% reduce 100%
13/12/09 22:34:07 INFO mapred.JobClient:     Job complete: job_201311240635_0196
13/12/09 22:34:07 INFO mapred.JobClient:     Counters: 29
13/12/09 22:34:07 INFO mapred.JobClient:     Job Counters
13/12/09 22:34:07 INFO mapred.JobClient:     Launched reduce tasks=1
13/12/09 22:34:07 INFO mapred.JobClient:     SLOTS_MILLIS_MAPS=8968
13/12/09 22:34:07 INFO mapred.JobClient:     Total time spent by all reduces waiting
after reserving slots (ms)=0
13/12/09 22:34:07 INFO mapred.JobClient:     Total time spent by all maps waiting after
reserving slots (ms)=0
13/12/09 22:34:07 INFO mapred.JobClient:     Launched map tasks=1
13/12/09 22:34:07 INFO mapred.JobClient:     SLOTS_MILLIS_REDUCES=10562
13/12/09 22:34:07 INFO mapred.JobClient:     File Output Format Counters
13/12/09 22:34:07 INFO mapred.JobClient:     Bytes Written=337623
13/12/09 22:34:07 INFO mapred.JobClient:     FileSystemCounters
13/12/09 22:34:07 INFO mapred.JobClient:     WASB_BYTES_READ=1395666
13/12/09 22:34:07 INFO mapred.JobClient:     FILE_BYTES_READ=466915
13/12/09 22:34:07 INFO mapred.JobClient:     HDFS_BYTES_READ=161
13/12/09 22:34:07 INFO mapred.JobClient:     FILE_BYTES_WRITTEN=1057448
13/12/09 22:34:07 INFO mapred.JobClient:     WASB_BYTES_WRITTEN=337623
13/12/09 22:34:07 INFO mapred.JobClient:     File Input Format Counters
13/12/09 22:34:07 INFO mapred.JobClient:     Bytes Read=1395667
13/12/09 22:34:07 INFO mapred.JobClient:     Map-Reduce Framework
13/12/09 22:34:07 INFO mapred.JobClient:     Map output materialized bytes=466761
13/12/09 22:34:07 INFO mapred.JobClient:     Map input records=32118
13/12/09 22:34:07 INFO mapred.JobClient:     Reduce shuffle bytes=466761
13/12/09 22:34:07 INFO mapred.JobClient:     Spilled Records=65912
13/12/09 22:34:07 INFO mapred.JobClient:     Map output bytes=2387798
13/12/09 22:34:07 INFO mapred.JobClient:     Total committed heap usage
(bytes)=1029046272
13/12/09 22:34:07 INFO mapred.JobClient:     CPU time spent (ms)=7420
13/12/09 22:34:07 INFO mapred.JobClient:     Combine input records=251357
13/12/09 22:34:07 INFO mapred.JobClient:     SPLIT_RAW_BYTES=161
13/12/09 22:34:07 INFO mapred.JobClient:     Reduce input records=32956
13/12/09 22:34:07 INFO mapred.JobClient:     Reduce input groups=32956
```

```
13/12/09 22:34:07 INFO mapred.JobClient:    Combine output records=32956
13/12/09 22:34:07 INFO mapred.JobClient:    Physical memory (bytes) snapshot=493834240
13/12/09 22:34:07 INFO mapred.JobClient:    Reduce output records=32956
13/12/09 22:34:07 INFO mapred.JobClient:    Virtual memory (bytes) snapshot=1430384640
13/12/09 22:34:07 INFO mapred.JobClient:    Map output records=251357
```

■ **Note**   The jobs you execute from the .NET and PowerShell programs are broken down internally as similar commands and executed as command-line jobs.

Make sure that the output files are created in the commandlineoutput folder as provided in the MapReduce command by issuing another ls command. This command lists the output file(s) created by the job as in Listing 6-4.

Listing 6-4. Verifying the output

```
c:\apps\dist\hadoop-1.2.0.1.3.1.0-06>hdfs fs -ls \example\data\commandlineoutput

Found 1 items -rw-r--r--    1 hadoopusersupergroup        337623 2013-12-09
22:34        /example/data/commandlineoutput/part-r-00000
```

You can copy output to the local file system and inspect the results (occurrences for each word will be in c:\output\part-r-00000) using the command in Listing 6-5.

Listing 6-5. Copying the MapReduce output from HDFS to local file system

```
hadoop dfs –copyToLocal /example/data/commandlineoutput c:\output
```

You can use Windows Explorer to view the output folder in your C:\Output directory as shown in Figure 6-8.

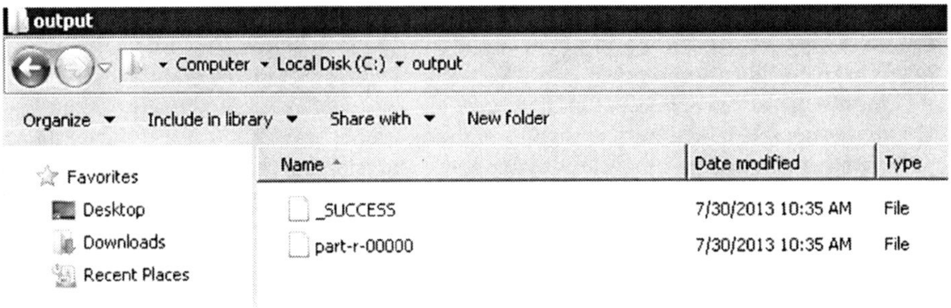

Figure 6-8. The output folder in the local file system

As indicated before, because Windows does not understand shell scripts for Linux (.sh files), all the command scripts and executables are implemented through Windows command files (.cmd files). You can use them directly from the command prompt as you would do in Linux, thus providing a complete abstraction to end users on Windows. For example, to start or stop your cluster, you can use the commands:

- stop-master.cmd

- stop-slave.cmd

Detailed descriptions of all the core Hadoop commands are beyond the scope of this book. If you are interested, you can refer to Apache's user manual on Hadoop commands for a complete listing and description at http://hadoop.apache.org/docs/r1.0.4/commands_manual.html.

A very important thing to re-iterate here is that the HDInsight Service actually simulates the HDFS behaviors for the end user. Actually, all the cluster data is stored in Windows Azure Storage Blob (WASB) in cluster-specific containers. If you remember, the core-site.xml file must have the entry of the Azure storage account and the account key to access the Azure blobs and function correctly. Here is the snippet of our cluster's core-site.xml, which uses the democluster blob as its cluster storage:

```
<property>
<name>fs.azure.account.key.democluster.blob.core.windows.net</name>
<value>******************************************************************* </value>
</property>
```

So the output folder and the file you just created is actually on your blob container for democluster. To confirm this, you can go to your Windows Azure Management Portal and see the blobs you just created as part of your cluster's data, as shown in Figure 6-9.

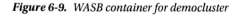

democlustercontai...	example/data/WordCountOutputPS/part-r-00000	http://democluster.bl(
	example/data/commandlineoutput	http://democluster.bl(
	example/data/commandlineoutput/part-r-00000	http://democluster.bl(
	example/data/gutenberg	http://democluster.bl(

Figure 6-9. *WASB container for democluster*

The Hive Console

Hive is an abstraction over HDFS and MapReduce. It enables you to define a table-like schema structure on the underlying HDFS (actually, WASB in HDInsight), and it provides a SQL-like query language to read data from the tables. The Hadoop Command Line also gives you access to the Hive console, from which you can directly execute the *Hive Query Language (HQL)*, to create, select, join, sort, and perform many other operations with the cluster data. Internally, the HQL queries are broken down to MapReduce jobs that execute and generate the desired output that is returned to the user. To launch the Hive console, navigate to the c:\apps\dist\hive-0.11.0.1.3.1.0-06\bin\ folder from the Hadoop Command Line and execute the Hive command. This should start the Hive command prompt as shown in Listing 6-6.

Listing 6-6. The Hive console

```
c:\apps\dist\hive-0.11.0.1.3.1.0-06\bin>hive

Logging initialized using configuration in file:/C:/apps/dist/hive-0.11.0.1.3.1.
0-06/conf/hive-log4j.properties
hive>
```

If you run the show tables command, it will show you similar output as you saw when you ran your Hive job from the .NET program in Chapter 5 as in Listing 6-7.

Listing 6-7. The show tables command

```
hive> show tables;
OK
aaplstockdata
hivesampletable
stock_analysis
stock_analysis1
Time taken: 3.182 seconds, Fetched: 4 row(s)
```

You can create new tables, populate them based on the data files in your blob containers in different partitions and query them based on different criteria directly from the Hive console. However, using .NET SDK and PowerShell are the recommended ways of making Hive job submissions in HDInsight rather than running them interactively from the console.

▪ **Note**   Details of Hive operations are covered in Chapter 8 of this book.

The Sqoop Console

Sqoop is an Open Source Apache project that facilitates bi-directional data exchange between Hadoop and any traditional *Relational Database Management System (RDBMS)*. It uses the MapReduce framework under the hood to perform import/export operations, and often it is a common choice for integrating data from relational and nonrelational data stores. In this section, we take a quick look at Sqoop operations that are compatible with Microsoft SQL Server on Azure.

Sqoop is based on *Java Database Connectivity (JDBC)* technology to establish connections to remote RDBMS servers. Therefore, you need the JDBC driver for SQL Server to be installed.

Table 6-1 summarizes a few of the key Sqoop operations that are supported with SQL Server databases in Azure.

Table 6-1. *Sqoop commands*

Command	Function
sqoop import	The import command lets you import SQL Server data into WABS. You can opt to import an entire table using the --table switch or selected records based on criteria using the --query switch. The data, once imported to the Azure storage system, is stored as delimited text files or as SequenceFiles for further processing. You can also use the import command to move SQL Server data into Hive tables, which are like logical schemas on top of WASB.
sqoop export	You can use the export command to move data from WASB into SQL Server tables. Much like the import command, the export command lets you export data from delimited text files, SequenceFiles, and Hive tables into SQL Server. The export command supports inserting new rows into the target SQL Server table, updating existing rows based on an update key column, as well as invoking a stored procedure execution.
sqoop job	The job command enables you to save your import/export commands as a job for future re-use. The saved jobs remember the parameters that are specified during execution, and they are particularly useful when there is a need to run an import or export command repeatedly on a periodic basis.
sqoop version	To quickly check the version of sqoop you are on, you can run the sqoop version command to print the installed version details on the console.

For example, assuming that you have a database called sqoopdemo deployed in SQL Azure that has a table called stock_analysis, you can execute the import command in Listing 6-8 to import that table's data into blob storage.

Listing 6-8. The Sqoop import command

```
sqoop import --connect "jdbc:sqlserver://<Server>.database.windows.net;username=debarchans@<Server>;
password=<Password>;database=sqoopdemo" --table stock_analysis --target-dir
example/data/StockAnalysis --as-textfile -m 1
```

On successful execution of the import job, you will see output on the Sqoop console similar to Listing 6-9.

Listing 6-9. The Sqoop import output

```
Warning: HBASE_HOME and HBASE_VERSION not set.
Warning: HBASE_HOME does not exist HBase imports will fail.
Please set HBASE_HOME to the root of your HBase installation.
13/12/10 01:04:42 INFO manager.SqlManager: Using default fetchSize of 1000
13/12/10 01:04:42 INFO tool.CodeGenTool: Beginning code generation

13/12/10 01:04:46 INFO manager.SqlManager: Executing SQL statement: SELECT t.* FROM
[stock_analysis] AS t WHERE 1=0
13/12/10 01:04:47 INFO orm.CompilationManager: HADOOP_MAPRED_HOME is
c:\apps\dist\hadoop-1.2.0.1.3.1.0-06
13/12/10 01:04:47 INFO orm.CompilationManager: Found hadoop core jar at: c:\apps\dist
\hadoop-1.2.0.1.3.1.0-06\hadoop-core.jar
Note: \tmp\sqoop-hadoopuser\compile\72c67877dd976aed8e4a36b3baa4519b\stock_analysis.java
uses or overrides a deprecated API.
Note: Recompile with -Xlint: deprecation for details.
13/12/10 01:04:49 INFO orm.CompilationManager: Writing jar file: \tmp\sqoop-hadoopuser\compile\72c67
877dd976aed8e4a36b3baa4519b\stock_analysis.jar
13/12/10 01:04:50 INFO mapreduce.ImportJobBase: Beginning import of stock_analysis

13/12/10 01:04:56 INFO mapred.JobClient:     Running job: job_201311240635_0197
13/12/10 01:04:57 INFO mapred.JobClient:     map 0% reduce 0%
13/12/10 01:05:42 INFO mapred.JobClient:     map 100% reduce 0%
13/12/10 01:05:45 INFO mapred.JobClient:     Job complete: job_201311240635_0197
13/12/10 01:05:45 INFO mapred.JobClient:     Counters: 19
13/12/10 01:05:45 INFO mapred.JobClient:     Job Counters
13/12/10 01:05:45 INFO mapred.JobClient:     SLOTS_MILLIS_MAPS=37452
13/12/10 01:05:45 INFO mapred.JobClient:     Total time spent by all reduces waiting after reserving
slots (ms)=0
13/12/10 01:05:45 INFO mapred.JobClient:     Total time spent by all maps waiting after
reserving slots (ms)=0
13/12/10 01:05:45 INFO mapred.JobClient:     Launched map tasks=1
13/12/10 01:05:45 INFO mapred.JobClient:     SLOTS_MILLIS_REDUCES=0
13/12/10 01:05:45 INFO mapred.JobClient:     File Output Format Counters
13/12/10 01:05:45 INFO mapred.JobClient:     Bytes Written=2148196
13/12/10 01:05:45 INFO mapred.JobClient:     FileSystemCounters
13/12/10 01:05:45 INFO mapred.JobClient:     FILE_BYTES_READ=770
13/12/10 01:05:45 INFO mapred.JobClient:     HDFS_BYTES_READ=87
13/12/10 01:05:45 INFO mapred.JobClient:     FILE_BYTES_WRITTEN=76307
13/12/10 01:05:45 INFO mapred.JobClient:     WASB_BYTES_WRITTEN=2148196
```

```
13/12/10 01:05:45 INFO mapred.JobClient:      File Input Format Counters
13/12/10 01:05:45 INFO mapred.JobClient:      Bytes Read=0
13/12/10 01:05:45 INFO mapred.JobClient:      Map-Reduce Framework
13/12/10 01:05:45 INFO mapred.JobClient:      Map input records=36153
13/12/10 01:05:45 INFO mapred.JobClient:      Physical memory (bytes) snapshot=215248896
13/12/10 01:05:45 INFO mapred.JobClient:      Spilled Records=0
13/12/10 01:05:45 INFO mapred.JobClient:      CPU time spent (ms)=5452
13/12/10 01:05:45 INFO mapred.JobClient:      Total committed heap usage (bytes)=514523136
13/12/10 01:05:45 INFO mapred.JobClient:      Virtual memory (bytes) snapshot=653586432
13/12/10 01:05:45 INFO mapred.JobClient:      Map output records=36153
13/12/10 01:05:45 INFO mapred.JobClient:      SPLIT_RAW_BYTES=87
```

**13/12/10 01:05:45 INFO mapreduce.ImportJobBase: Transferred 0 bytes in 54.0554
seconds (0 bytes/sec)**
13/12/10 01:05:45 INFO mapreduce.ImportJobBase: Retrieved 36153 records.

Windows PowerShell also provides cmdlets to execute Sqoop jobs. The following PowerShell script in Listing 6-10 exports the same StockAnalysis blob from WASB to a SQL Azure database called ExportedData.

Listing 6-10. The Sqoop export PowerShell script

```
$subscriptionName= "Your_Subscription_Name"
$clusterName = "democluster"
$SqoopCommand = "export -connect
`"jdbc:sqlserver://<Server>.database.windows.net;username=debarchans@<Server>;
    password=<Password>;database=sqoopdemo`" --table stock_analysis
    --export-dir /user/hadoopuser/example/data/StockAnalysis
    --input-fields-terminated-by `","`""
$sqoop = New-AzureHDInsightSqoopJobDefinition -Command $SqoopCommand
$SqoopJob = Start-AzureHDInsightJob -Subscription (Get-AzureSubscription
  -Current).SubscriptionId -Cluster $clustername -JobDefinition $sqoop
Wait-AzureHDInsightJob -Subscription (Get-AzureSubscription
  -Current).SubscriptionId -Job $SqoopJob -WaitTimeoutInSeconds 3600
Get-AzureHDInsightJobOutput -Cluster $clusterName -JobId $SqoopJob.JobId
  -StandardError -Subscription $subscriptionName
```

Successful execution of the Sqoop export job shows output similar to Listing 6-11.

Listing 6-11. The PowerShell Sqoop export output

```
StatusDirectory : ee84101c-98ac-4a2b-ae3d-49600eb5954b
ExitCode        : 0
Name            :

Query           : export --connect
"jdbc:sqlserver://<Server>.database.windows.net;username=debarchans@<Server>;
password=<Password>;database=sqoopdemo" --table stock_analysis --export-dir
/user/hadoopuser/example/data/StockAnalysis --input-fields-terminated-by ","
State           : Completed
SubmissionTime  : 12/10/2013 1:36:36 AM
Cluster         : democluster
PercentComplete : map 100% reduce 0%
JobId           : job_201311240635_0205
```

```
D:\python27\python.exe: can't open file '\bin\hcat.py': [Errno 2] No such
file or directory
13/12/10 01:36:48 INFO manager.SqlManager: Using default fetchSize of 1000
13/12/10 01:36:48 INFO tool.CodeGenTool: Beginning code generation
13/12/10 01:36:52 INFO manager.SqlManager: Executing SQL statement: SELECT t.*
FROM [stock_analysis] AS t WHERE 1=0
13/12/10 01:36:53 INFO orm.CompilationManager: HADOOP_MAPRED_HOME is c:\hdfs\mapred\local\
taskTracker\admin\jobcache\job_201311240635_0205\attempt_
201311240635_0205_m_000000_0\work\"C:\apps\dist\hadoop-1.2.0.1.3.1.0-06"
13/12/10 01:36:53 WARN orm.CompilationManager: HADOOP_MAPRED_HOME appears empty
or missing
Note: \tmp\sqoop-hdp\compile\c2070a7782f921c6cd0cfd58ab7efe66\stock_analysis.java
uses or overrides a deprecated API.
Note: Recompile with -Xlint:deprecation for details.
13/12/10 01:36:54 INFO orm.CompilationManager: Writing jar file: \tmp\sqoop-
hdp\compile\c2070a7782f921c6cd0cfd58ab7efe66\stock_analysis.jar
13/12/10 01:36:54 INFO mapreduce.ExportJobBase: Beginning export of stock_analysis
13/12/10 01:36:58 INFO input.FileInputFormat: Total input paths to process : 1
13/12/10 01:36:58 INFO input.FileInputFormat: Total input paths to process : 1
13/12/10 01:36:58 WARN snappy.LoadSnappy: Snappy native library is available
13/12/10 01:36:58 INFO util.NativeCodeLoader: Loaded the native-hadoop library
13/12/10 01:36:58 INFO snappy.LoadSnappy: Snappy native library loaded
13/12/10 01:36:58 INFO mapred.JobClient: Running job: job_201311240635_0206
13/12/10 01:37:00 INFO mapred.JobClient:  map 0% reduce 0%
13/12/10 01:37:21 INFO mapred.JobClient:  map 10% reduce 0%
13/12/10 01:37:22 INFO mapred.JobClient:  map 16% reduce 0%
13/12/10 01:37:23 INFO mapred.JobClient:  map 21% reduce 0%
13/12/10 01:37:27 INFO mapred.JobClient:  map 27% reduce 0%
13/12/10 01:37:28 INFO mapred.JobClient:  map 32% reduce 0%
13/12/10 01:37:30 INFO mapred.JobClient:  map 41% reduce 0%
13/12/10 01:37:33 INFO mapred.JobClient:  map 46% reduce 0%
13/12/10 01:37:34 INFO mapred.JobClient:  map 55% reduce 0%
13/12/10 01:37:35 INFO mapred.JobClient:  map 63% reduce 0%
13/12/10 01:37:36 INFO mapred.JobClient:  map 71% reduce 0%
13/12/10 01:37:37 INFO mapred.JobClient:  map 77% reduce 0%
13/12/10 01:37:39 INFO mapred.JobClient:  map 82% reduce 0%
13/12/10 01:37:40 INFO mapred.JobClient:  map 85% reduce 0%
13/12/10 01:37:41 INFO mapred.JobClient:  map 88% reduce 0%
13/12/10 01:37:42 INFO mapred.JobClient:  map 94% reduce 0%
13/12/10 01:37:43 INFO mapred.JobClient:  map 99% reduce 0%
13/12/10 01:37:45 INFO mapred.JobClient:  map 100% reduce 0%
13/12/10 01:37:50 INFO mapred.JobClient: Job complete: job_201311240635_0206
13/12/10 01:37:50 INFO mapred.JobClient: Counters: 20
13/12/10 01:37:50 INFO mapred.JobClient:   Job Counters
13/12/10 01:37:50 INFO mapred.JobClient:     SLOTS_MILLIS_MAPS=151262
13/12/10 01:37:50 INFO mapred.JobClient:    Total time spent by all reduces
waiting after reserving slots (ms)=0
13/12/10 01:37:50 INFO mapred.JobClient:    Total time spent by all maps waiting
after reserving slots (ms)=0
13/12/10 01:37:50 INFO mapred.JobClient:    Rack-local map tasks=4
13/12/10 01:37:50 INFO mapred.JobClient:    Launched map tasks=4
```

```
13/12/10 01:37:50 INFO mapred.JobClient:       SLOTS_MILLIS_REDUCES=0
13/12/10 01:37:50 INFO mapred.JobClient:     File Output Format Counters
13/12/10 01:37:50 INFO mapred.JobClient:       Bytes Written=0
13/12/10 01:37:50 INFO mapred.JobClient:     FileSystemCounters
13/12/10 01:37:50 INFO mapred.JobClient:       WASB_BYTES_READ=3027416
13/12/10 01:37:50 INFO mapred.JobClient:       FILE_BYTES_READ=3696
13/12/10 01:37:50 INFO mapred.JobClient:       HDFS_BYTES_READ=792
13/12/10 01:37:50 INFO mapred.JobClient:       FILE_BYTES_WRITTEN=296608
13/12/10 01:37:50 INFO mapred.JobClient:     File Input Format Counters
13/12/10 01:37:50 INFO mapred.JobClient:       Bytes Read=0
13/12/10 01:37:50 INFO mapred.JobClient:     Map-Reduce Framework
13/12/10 01:37:50 INFO mapred.JobClient:       Map input records=36153
13/12/10 01:37:50 INFO mapred.JobClient:       Physical memory (bytes) snapshot=779915264
13/12/10 01:37:50 INFO mapred.JobClient:       Spilled Records=0
13/12/10 01:37:50 INFO mapred.JobClient:       CPU time spent (ms)=17259
13/12/10 01:37:50 INFO mapred.JobClient:       Total committed heap usage (bytes)=2058092544
13/12/10 01:37:50 INFO mapred.JobClient:       Virtual memory (bytes) snapshot=2608484352
13/12/10 01:37:50 INFO mapred.JobClient:       Map output records=36153
13/12/10 01:37:50 INFO mapred.JobClient:       SPLIT_RAW_BYTES=792
13/12/10 01:37:50 INFO mapreduce.ExportJobBase:Transferred 792 bytes in 53.6492
seconds (14.7626 bytes/sec)
13/12/10 01:37:50 INFO mapreduce.ExportJobBase: Exported 36153 records.
```

As you can see, Sqoop is a pretty handy import/export tool for your cluster's data, allowing you to go easily to and from a SQL Azure database. Sqoop allows you to merge structured and unstructured data, and to provide powerful analytics on the data overall. For a complete reference of all the available Sqoop commands, visit the Apache documentation site at https://cwiki.apache.org/confluence/display/SQOOP/Home.

The Pig Console

Pig is a set-based *data transformation* tool that works on top of the Hadoop stack to manipulate data sets to add and remove aggregates, and to transform data. Pig is most analogous to the Dataflow task in *SQL Server Integration Services (SSIS)*, as discussed in Chapter 10.

Unlike SSIS, Pig does not have a control-flow system. Pig is written in Java and produces Java `.jar` code to run MapReduce jobs across the nodes in the Hadoop cluster to manipulate the data in a distributed way. Pig exposes a command-line shell called Grunt to execute Pig statements. To launch the Grunt shell, navigate to `c:\apps\dist\pig-0.11.0.1.3.1.0-06\bin` directory from the Hadoop Command Line. Then execute the `Pig` command. That should launch the Grunt shell as shown in Listing 6-12.

Listing 6-12. Launching the Pig Grunt shell

```
c:\apps\dist\pig-0.11.0.1.3.1.0-06\bin>pig
2013 -12-10 01:48:10,150 [main] INFO  org.apache.pig.Main - Apache Pig version 0.11.0.1.3.1.0-06
(r: unknown) compiled Oct 02 2013, 21:58:30
2013 -12-10 01:48:10,151 [main] INFO  org.apache.pig.Main - Logging error messages to:
C:\apps\dist\hadoop-1.2.0.1.3.1.0-06\logs\pig_1386640090147.log
2013 -12-10 01:48:10,194 [main] INFO  org.apache.pig.impl.util.Utils
- Default bootup file D:\Users\hadoopuser/.pigbootup not found
2013 -12-10 01:48:10,513 [main] INFO  org.apache.pig.backend.hadoop.executionengine.HExecutionEngine
- Connecting to hadoop file system at: wasb://democlustercontainer@democluster.blob.core.windows.net
```

```
2013 -12-10 01:48:11,279 [main] INFO  org.apache.pig.backend.hadoop.executionengine.HExecutionEngine
- Connecting to map-reduce job tracker at: jobtrackerhost:9010
grunt>
```

Let's execute a series of Pig statements to parse the Sample.log file that is present in the /example/data/folder by default in WASB containers. The first statement loads the file content to a Pig variable called LOGS:

```
LOGS = LOAD 'wasb:///example/data/sample.log';
```

Then we will create a variable LEVELS that will categorize the entries in the LOGS variable based on Info, Error, Warnings, and so forth. For example:

```
LEVELS = foreach LOGS generate REGEX_EXTRACT($0,'(TRACE|DEBUG|INFO|WARN|ERROR|FATAL)'
, 1)  as LOGLEVEL;
```

Next, we can filter out the null entries in the FILTEREDLEVEL variables:

```
FILTEREDLEVELS = FILTER LEVELS by LOGLEVEL is not null;
```

After that, we can filter the group entries based on the values in the variable GROUPEDLEVELS:

```
GROUPEDLEVELS = GROUP FILTEREDLEVELS by LOGLEVEL;
```

Next, we count the number of occurrences of each entry type and load them in the FREQUENCIES variable. For example:

```
FREQUENCIES = foreach GROUPEDLEVELS generate group as LOGLEVEL,
COUNT(FILTEREDLEVELS.LOGLEVEL) as COUNT;
```

Then we arrange the grouped entries in descending order of their number of occurrences in the RESULTS variable. Here's how to sort in that order:

```
RESULT = order FREQUENCIES by COUNT desc;
```

Finally, we can print out the value of the RESULTS variable using the DUMP command. Note that this is the place where the actual MapReduce job is triggered to process and fetch the data. Here's the command:

```
DUMP RESULT;
```

On successful execution of Pig statements, you should see output where the log entries are grouped by their values and arranged based on their number of occurrences. Such output is shown in Listing 6-13.

Listing 6-13. The Pig job output

```
Input(s):
Successfully read 1387 records (404 bytes) from: "wasb:///example/data/sample.log"

Output(s):
Successfully stored 6 records in: "wasb://democlustercontainer@democluster.blob.
core.windows.net/tmp/temp167788958/tmp-1711466614"
```

```
Counters:
Total records written : 6
Total bytes written : 0
Spillable Memory Manager spill count : 0
Total bags proactively spilled: 0
Total records proactively spilled: 0

Job DAG:
job_201311240635_0221    ->      job_201311240635_0222,
job_201311240635_0222    ->      job_201311240635_0223,
job_201311240635_0223


2013-12-10 02:24:01,797 [main] INFO  org.apache.pig.backend.hadoop.executionengine.mapReduceLayer.
MapReduceLauncher - Success!
2013-12-10 02:24:01,800 [main] INFO  org.apache.pig.data.SchemaTupleBackend - Key [pig.schematuple]
was not set... will not generate code.
2013-12-10 02:24:01,825 [main] INFO  org.apache.hadoop.mapreduce.lib.input.FileInputFormat - Total
input paths to process : 1
2013-12-10 02:24:01,825 [main] INFO  org.apache.pig.backend.hadoop.executionengine.util.MapRedUtil -
Total input paths to process : 1


(TRACE,816)
(DEBUG,434)
(INFO,96)
(WARN,11)
(ERROR,6)
(FATAL,2)
```

Much like all the other types of jobs, Pig jobs can also be submitted using a PowerShell script. Listing 6-14 shows the PowerShell script to execute the same Pig job.

Listing 6-14. The PowerShell Pig job

```
$subid = "Your_Subscription_Id"
$subName = "your_Subscription_name"
$clusterName = "democluster"
$0 = '$0';
$QueryString= "LOGS = LOAD 'wasb:///example/data/sample.log';" +
"LEVELS = foreach LOGS generate REGEX_EXTRACT($0, '(TRACE|DEBUG|INFO|WARN|ERROR|FATAL)',
1) as LOGLEVEL;" +
"FILTEREDLEVELS = FILTER LEVELS by LOGLEVEL is not null;" +
"GROUPEDLEVELS = GROUP FILTEREDLEVELS by LOGLEVEL;" +
"FREQUENCIES = foreach GROUPEDLEVELS generate group as LOGLEVEL, COUNT(FILTEREDLEVELS.LOGLEVEL)
as COUNT;" +
"RESULT = order FREQUENCIES by COUNT desc;" +
"DUMP RESULT;"


$pigJobDefinition = New-AzureHDInsightPigJobDefinition -Query $QueryString -StatusFolder
"/PigJobs/PigJobStatus"
#Submit the Pig Job to the cluster
$pigJob = Start-AzureHDInsightJob -Subscription $subid -Cluster $clusterName -JobDefinition
$pigJobDefinition
```

```
#Wait for the job to complete
$pigJob | Wait-AzureHDInsightJob -Subscription $subid -WaitTimeoutInSeconds 3600
```

Using the Grunt shell in Pig is another way to bypass coding MapReduce jobs, which can be tedious and time consuming. The HDInsight name node gives you the option to interactively run Pig commands from their respective command shells. Doing so is often a great way to isolate any external issues when you are facing errors while submitting jobs using PowerShell or .NET.

Hadoop Web Interfaces

Core Hadoop provides a couple of web interfaces to monitor your cluster and, by default, they are available at the desktop of the name node. These portals can provide useful details about the cluster health, usage and MapReduce job execution statistics. The shortcuts to these portals are created on the desktop during the Azure virtual machine (VM) provisioning process, as shown in Figure 6-10. They are

- Hadoop MapReduce Status
- Hadoop Name Node Status

Figure 6-10. *Shortcuts to the web portals*

Hadoop MapReduce Status

The Hadoop MapReduce portal displays information on job configuration parameters and execution statistics in terms of running/completed/failed jobs. The portal also shows job history log files. You can drill down on individual jobs and examine the details.

The portal is referred to as the *JobTracker* portal, because each MapReduce operation is submitted and executed as a job in the cluster. The tracker portion of the portal is basically a Java-based web application that listens on port 50030.

The URL for the portal is http://<NameNode_IP_Address>:50030/jobtracker.jsp. Figure 6-11 shows the MapReduce status or the JobTracker status portal when it is launched.

jobtrackerhost Hadoop Map/Reduce Administration

State: RUNNING
Started: Tue Dec 10 02:47:00 GMT 2013
Version: 1.2.0.1.3.1.0-06, rf4cb3bb77cf3cc20c863de73bd6ef21cf069f66f
Compiled: Wed Oct 02 21:38:25 Coordinated Universal Time 2013 by jenkins
Identifier: 201312100246
SafeMode: OFF

Cluster Summary (Heap Size is 616.38 MB/3.56 GB)

Running Map Tasks	Running Reduce Tasks	Total Submissions	Nodes	Occupied Map Slots	Occupied Reduce Slots	Reserved Map Slots	Reserved Reduce Slots	Map Task Capacity	Reduce Task Capacity	Avg. Tasks/Node
0	0	14	2	0	0	0	0	8	4	6.00

Figure 6-11. The MapReduce Status portal

You can scroll down to see the list of completed jobs, running jobs (which would populate only if a job is running at that point), failed jobs, and retired jobs. You can click on any of the job records to view more details about that specific operation as shown in Figure 6-12.

Status: Succeeded
Started at: Tue Dec 10 03:53:54 GMT 2013
Finished at: Tue Dec 10 03:54:19 GMT 2013
Finished in: 25sec
Job Cleanup: Successful
Job Scheduling information: 0 running map tasks using 0 map slots. 0 additional slots reserved. 0 running reduce tasks using 0

Kind	% Complete	Num Tasks	Pending	Running	Complete	Killed	Failed/Killed Task Attempts
map	100.00%	1	0	0	1	0	0 / 0
reduce	100.00%	0	0	0	0	0	0 / 0

	Counter	Map	Reduce	Total
Job Counters	SLOTS_MILLIS_MAPS	0	0	22,047
	Total time spent by all reduces waiting after reserving slots (ms)	0	0	0
	Total time spent by all maps waiting after reserving slots (ms)	0	0	0
	Launched map tasks	0	0	1
	SLOTS_MILLIS_REDUCES	0	0	0
File Output Format Counters	Bytes Written	0	0	0
File Input Format Counters	Bytes Read	0	0	0
FileSystemCounters	WASB_BYTES_READ	164	0	164
	FILE_BYTES_READ	462	0	462
	HDFS_BYTES_READ	45	0	45
	FILE_BYTES_WRITTEN	63,851	0	63,851
	WASB_BYTES_WRITTEN	238	0	238

Figure 6-12. MapReduce job statistics

The Hadoop MapReduce portal gives you a comprehensive summary on each of the submitted jobs. You can drill down into the stdout and stderr output of the jobs, so it is obvious that the portal is a great place to start troubleshooting a MapReduce job problem.

The Name Node Status Portal

The Hadoop Name Node Status web interface shows a cluster summary, including information about total and remaining capacity, the file system, and cluster health. The interface also gives the number of live, dead, and decommissioning nodes. The Name Node Status Portal is a Java web application that listens on port 50070. It can be launched from the URL http://<NameNode_IP_Address>:50070/dfshealth.jsp.

Additionally, the Name Node Status Portal allows you to browse the HDFS (actually, WASB) namespace and view the contents of its files in the web browser. It also gives access to the name node's log files for Hadoop services. At a glance, this portal gives you an overview of how your cluster is doing, as shown in Figure 6-13.

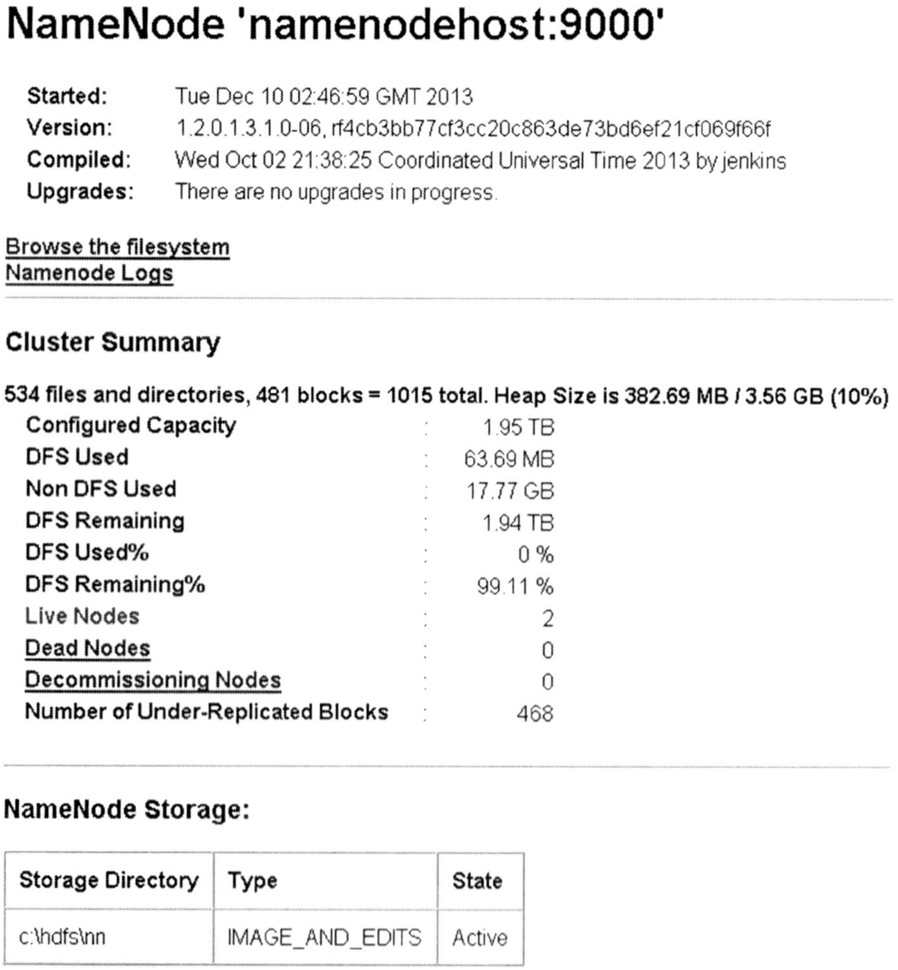

NameNode 'namenodehost:9000'

Started:	Tue Dec 10 02:46:59 GMT 2013
Version:	1.2.0.1.3.1.0-06, rf4cb3bb77cf3cc20c863de73bd6ef21cf069f66f
Compiled:	Wed Oct 02 21:38:25 Coordinated Universal Time 2013 by jenkins
Upgrades:	There are no upgrades in progress.

Browse the filesystem
Namenode Logs

Cluster Summary

534 files and directories, 481 blocks = 1015 total. Heap Size is 382.69 MB / 3.56 GB (10%)

Configured Capacity	:	1.95 TB
DFS Used	:	63.69 MB
Non DFS Used	:	17.77 GB
DFS Remaining	:	1.94 TB
DFS Used%	:	0 %
DFS Remaining%	:	99.11 %
Live Nodes	:	2
Dead Nodes	:	0
Decommissioning Nodes	:	0
Number of Under-Replicated Blocks	:	468

NameNode Storage:

Storage Directory	Type	State
c:\hdfs\nn	IMAGE_AND_EDITS	Active

Figure 6-13. *The Name Node Status Portal*

You can drill down on the data nodes, access their file system, and go all the way to job configurations used during job submission, as shown in Figure 6-14.

Contents of directory /mapred/userhistory/_logs/history

Goto : /mapred/userhistory/_logs [go]

Go to parent directory

Name	Type	Size	Replication	Block Size	Modification Time	Permission	Owner
job_201311240635_0001_1385276716108_admin_TempletonControllerJob_joblauncher_%20F	file	9.1 KB	3	256 MB	2013-11-24 07:05	rw-r--r--	admin
job_201311240635_0001_conf.xml	file	53.68 KB	3	256 MB	2013-11-24 07:05	rw-r--r--	admin
job_201311240635_0002_1385276905889_admin_TempletonControllerJob_joblauncher_%20F	file	9.27 KB	3	256 MB	2013-11-24 07:09	rw-r--r--	admin
job_201311240635_0002_conf.xml	file	53.58 KB	3	256 MB	2013-11-24 07:08	rw-r--r--	admin
job_201311240635_0003_1385277078913_admin_TempletonControllerJob_joblauncher_%20F	file	9.27 KB	3	256 MB	2013-11-24 07:11	rw-r--r--	admin
job_201311240635_0003_conf.xml	file	53.46 KB	3	256 MB	2013-11-24 07:11	rw-r--r--	admin
job_201311240635_0004_1385277772398_admin_TempletonControllerJob_joblauncher_%20F	file	9.27 KB	3	256 MB	2013-11-24 07:23	rw-r--r--	admin

Figure 6-14. *The job configurations*

The Name Node Status Portal is a part of the Apache Hadoop project, making it familiar to existing Hadoop users. The main advantage of the portal is that it lets you browse through the file system as if it is a local file system. That's an advantage because there is no way to access the file system through standard tools like Windows Explorer, as the entire storage mechanism is abstracted in WASB.

The TaskTracker Portal

Apart from the Name Node and MapReduce status portals, there is also a TaskTracker web interface that is available only in the data nodes or task nodes of your cluster. This portal listens on port 50060, and the complete URL to launch it is http://<DataNode_IP_Address>:50060/tasktracker.jsp. Although there is a single TaskTracker per slave node, each TaskTracker can spawn multiple JVMs to handle many map or reduce tasks in parallel.

■ **Note** The TaskTracker service runs on the data nodes, so there is no shortcut created for that portal in the name node.

You need to log on remotely to any of your data nodes to launch the TaskTracker portal. Remember, the remote logon session needs to be initiated from the name node Remote Desktop session itself. It will not work if you try to connect remotely to your data node from your client workstation. This Java-based web portal displays the status of the completed tasks along with their status, as shown in Figure 6-15.

Running tasks

Task Attempts	Status	Progress	Errors

Non-Running Tasks

Task Attempts	Status
attempt_201312100246_0021_m_000001_0	SUCCEEDED
attempt_201312100246_0021_m_000000_0	SUCCEEDED

Tasks from Running Jobs

Task Attempts	Status	Progress	Errors
attempt_201312100246_0021_m_000001_0	SUCCEEDED	100.00%	
attempt_201312100246_0021_m_000000_0	SUCCEEDED	100.00%	

Figure 6-15. *The TaskTracker web portal*

The Running tasks section of the TaskTracker is populated only if a job (which comprises one or more tasks) is in execution at that point of time. If any MapReduce job is running in the cluster, this section will show the details of each of the Map and Reduce tasks, as shown in Figure 6-16.

Running tasks

Task Attempts	Status	Progress	Errors
attempt_201312100246_0025_m_000000_0	RUNNING	0.00%	
attempt_201312100246_0027_r_000000_0	COMMIT_PENDING	33.33%	

Figure 6-16. *The running tasks in TaskTracker*

While the JobTracker or the MapReduce service tracker is the master monitoring the overall execution of a MapReduce job, the TaskTrackers manage the execution of individual tasks on each slave node. Another important responsibility of the TaskTracker is to constantly communicate with the JobTracker. If the JobTracker fails to receive a heartbeat from a TaskTracker within a specified amount of time, it will assume the TaskTracker has crashed and will resubmit the corresponding tasks to other nodes in the cluster.

HDInsight Windows Services

In traditional Hadoop, each process like the namenode, datanode, and so on are known as *daemons*, which stands for *Disk and Execution Monitor*. In simple terms, a daemon is a long-running background process that answers requests for services. In the Windows environment, they are called *services*. Windows provides a centralized way to view and manage services running in the system through a console known as the Services console. Hadoop daemons are translated to Windows services in HDInsight distribution. To view the Hadoop services running on your cluster head node, click on Start ➤ Run and type in Services.msc. This will launch the Services console, and you will see the different Apache Hadoop-related services, as shown in Figure 6-17.

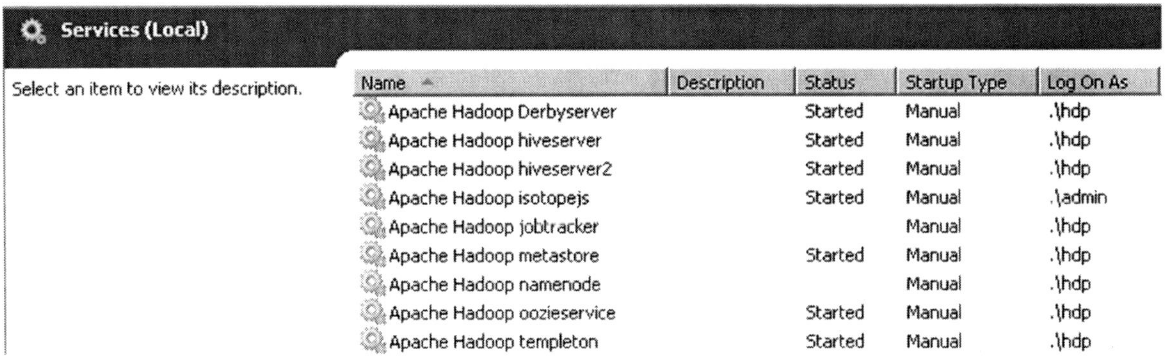

Name ▲	Description	Status	Startup Type	Log On As
Apache Hadoop Derbyserver		Started	Manual	.\hdp
Apache Hadoop hiveserver		Started	Manual	.\hdp
Apache Hadoop hiveserver2		Started	Manual	.\hdp
Apache Hadoop isotopejs		Started	Manual	.\admin
Apache Hadoop jobtracker			Manual	.\hdp
Apache Hadoop metastore		Started	Manual	.\hdp
Apache Hadoop namenode			Manual	.\hdp
Apache Hadoop oozieservice		Started	Manual	.\hdp
Apache Hadoop templeton		Started	Manual	.\hdp

Figure 6-17. *Apache Hadoop Windows services*

These services are unique to Hadoop on Windows, and Table 6-2 summarizes the function of each service.

Table 6-2. *The Hadoop Windows services*

Service	Function
Apache Hadoop Derbyserver	Runs the service for Hive's native embedded database technology called *Derby*
Apache Hadoop hiveserver	Simulates Hive's native thrift service for remote client connectivity
Apache Hadoop hiveserver2	Same as hiveserver with support for concurrency for ODBC and JDBC
Apache Hadoop isotopejs	Runs the required handlers for the interactive consoles that are available on the HDInsight management portal
Apache Hadoop jobtracker	Runs the Hadoop job tracker service
Apache Hadoop metastore	Runs the Hive/Oozie metastore services
Apache Hadoop namenode	Runs the Hadoop NameNode service
Apache Hadoop oozieservice	Runs the Oozie service
Apache Hadoop templeton	Runs the Templeton service

Access to the services in Table 6-2 gives you control of the different programs you need to run on your Hadoop cluster. If it is a really busy cluster doing only core MapReduce processing, you might want to stop the services for a few supporting projects like Hive and Oozie, which are not used at that point.

Your Azure Management portal gives you an option to turn all Hadoop services on or off as a whole, as shown in Figure 6-18. However, through the name node's Services console, you can selectively turn off or on any of the services you want.

democluster
DASHBOARD
MONITOR
CONFIGURATION
SDPHDI1
democluster
datadork
tutorial
cluster connectivity
HADOOP SERVICES
ON
OFF

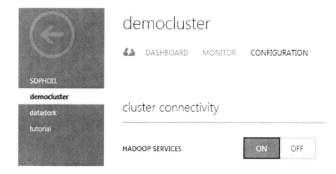

Figure 6-18. *Toggle Hadoop services*

Installation Directory

HDInsight distribution deploys core Hadoop and the supporting projects to the `C:\apps\dist` directory of the name node. The folder and directory structure of the components are almost the same as in the Open Source projects to maintain consistency and compatibility. The directory structure for your name node should look like Figure 6-19.

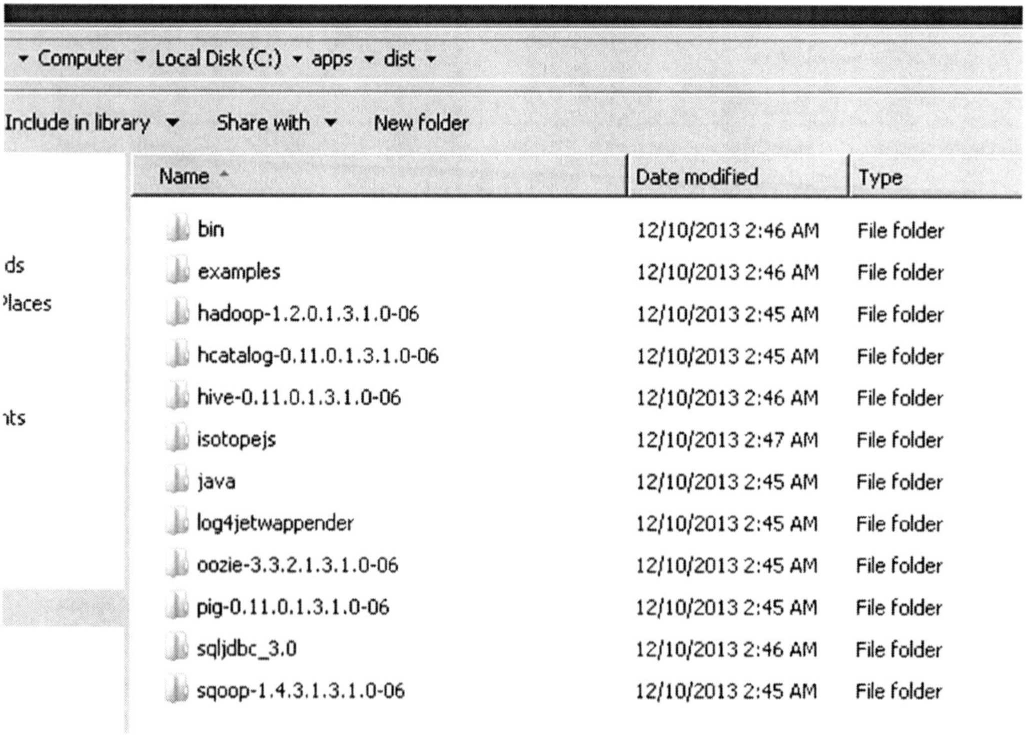
Computer ▸ Local Disk (C:) ▸ apps ▸ dist
Include in library
Share with
New folder
Name
Date modified
Type
bin 12/10/2013 2:46 AM File folder
examples 12/10/2013 2:46 AM File folder
hadoop-1.2.0.1.3.1.0-06 12/10/2013 2:45 AM File folder
hcatalog-0.11.0.1.3.1.0-06 12/10/2013 2:45 AM File folder
hive-0.11.0.1.3.1.0-06 12/10/2013 2:46 AM File folder
isotopejs 12/10/2013 2:47 AM File folder
java 12/10/2013 2:45 AM File folder
log4jetwappender 12/10/2013 2:45 AM File folder
oozie-3.3.2.1.3.1.0-06 12/10/2013 2:45 AM File folder
pig-0.11.0.1.3.1.0-06 12/10/2013 2:45 AM File folder
sqljdbc_3.0 12/10/2013 2:46 AM File folder
sqoop-1.4.3.1.3.1.0-06 12/10/2013 2:45 AM File folder

Figure 6-19. *Hadoop on the Windows installation directory*

Note The Java runtime is also deployed in the same directory.

Summary

In this chapter, you read about enabling Remote Desktop and logging on to the HDInsight cluster's name node with proper cluster credentials. The name node is the heart of the cluster, and you can do all the operations from the name node that you can from the management portal or the .NET SDK and PowerShell scripts.

The name node gives you access to the Hadoop command line and the web interfaces that are available with the distribution. HDInsight simulates WASB as HDFS behind the scenes for the end users. You saw how actually all the input and output files are saved back to your Azure storage account dedicated for the cluster through the Azure Management portal. The WASB mechanism is an abstraction to the user, who sees a simulation of HDFS when dealing with file system operations. You learned to execute basic HDFS/MapReduce commands using the command line and about the different unique Windows services for Hadoop. You also had a look at the different supporting projects like Hive, Sqoop, and Pig and how they can be invoked from the command line as well as from PowerShell scripts. Finally, we navigated through the installation files and folder hierarchies of Hadoop and the other projects in the C:\apps\ dist directory of the name node.

CHAPTER 7

Using Windows Azure HDInsight Emulator

Deploying your Hadoop clusters on Azure invariably incurs some cost. The actual cost of deploying a solution depends on the size of your cluster, the data you play with, and certain other aspects, but there are some bare-minimum expenses for even setting up a test deployment for evaluation. For example, you will at least need to pay for your Azure subscription in order to try the HDInsight service on Azure. This is not acceptable for many individuals or institutions who want to evaluate the technology and then decide on an actual implementation. Also, you need to have a test bed to test your solutions before deploying them to an actual production environment. To address these scenarios, Microsoft offers the Windows Azure HDInsight Emulator.

The Windows Azure HDInsight Emulator is an implementation of HDInsight on the Windows Server family. The emulator is currently available as a Developer Preview, where the Hadoop-based services on Windows use only a single-node deployment. HDInsight Emulator provides you with a local development environment for the Windows Azure HDInsight Service. It uses the same software bits as the Azure HDInsight service and is the test bed recommended by Microsoft for testing and evaluation.

Caution   While it's technically possible create a multinode configuration of HDInsight emulator, doing so is neither a recommended nor a supported scenario, because it opens the door to serious security breaches in your environment. If you are still eager to do the multinode configuration and you delete the firewall rule and modify the `*-conf.xml` Hadoop config files, you'll essentially be allowing anyone to run code on your machine and access your file system. However, such a configuration can be tested in a less sensitive lab environment solely for testing purposes and is documented in the following blog post: `http://binyoga.blogspot.in/2013/07/virtual-lab-multi-node-hadoop-cluster.html`.

Like the Azure service, the emulator is also based on Hortonworks Data Platform (HDP), which bundles all the Apache projects under the hood and makes it compatible with Windows. This local development environment for HDInsight simplifies the configuration, execution, and processing of Hadoop jobs by providing a PowerShell library with HDInsight cmdlets for managing the cluster and the jobs run on it. It also provides a .NET SDK for HDInsight for automating these procedures—again, much like the Azure service. For users who need multinode Hadoop solutions on their on-premises Windows servers today, the recommended option is to use HDP for Windows. Microsoft has no plans whatsoever to make this emulator multinode and give it the shape of a production on-premises Hadoop cluster on Windows.

Installing the Emulator

The Windows Azure HDInsight Emulator is installed with the Microsoft Web Platform Installer version 4.5 or higher. The current distribution of HDInsight Emulator installs HDP 1.1 for Windows. For more details about different HDP versions, visit the Hortonworks web site:

```
http://hortonworks.com/products/hdp/
```

■ **Note** The Microsoft Web Platform Installer (Web PI) is a free tool that makes getting the latest components of the Microsoft Web Platform—including Internet Information Services (IIS), SQL Server Express, .NET Framework, and Visual Web Developer—easy. The Web PI also makes it easy to install and run the most popular free web applications for blogging, content management, and more with the built-in Windows Web Application Gallery.

HDP 1.1 includes HDInsight cluster version 1.6. Microsoft plans to upgrade the emulator and match the version (which, as of now, is version 2.1) that is deployed in the Azure Service.

The emulator currently supports Windows 7, Windows 8, and the Windows Server 2012 family of operating systems. It can be downloaded from the following link:

```
http://www.microsoft.com/web/handlers/webpi.ashx/getinstaller/HDINSIGHT-PREVIEW.appids
```

You can also go to the Emulator installation page and launch the installer:

```
http://www.microsoft.com/web/gallery/install.aspx?appid=HDINSIGHT
```

When prompted, execute the installer and you should see the Web Platform Installer ready to install the emulator as shown in Figure 7-1.

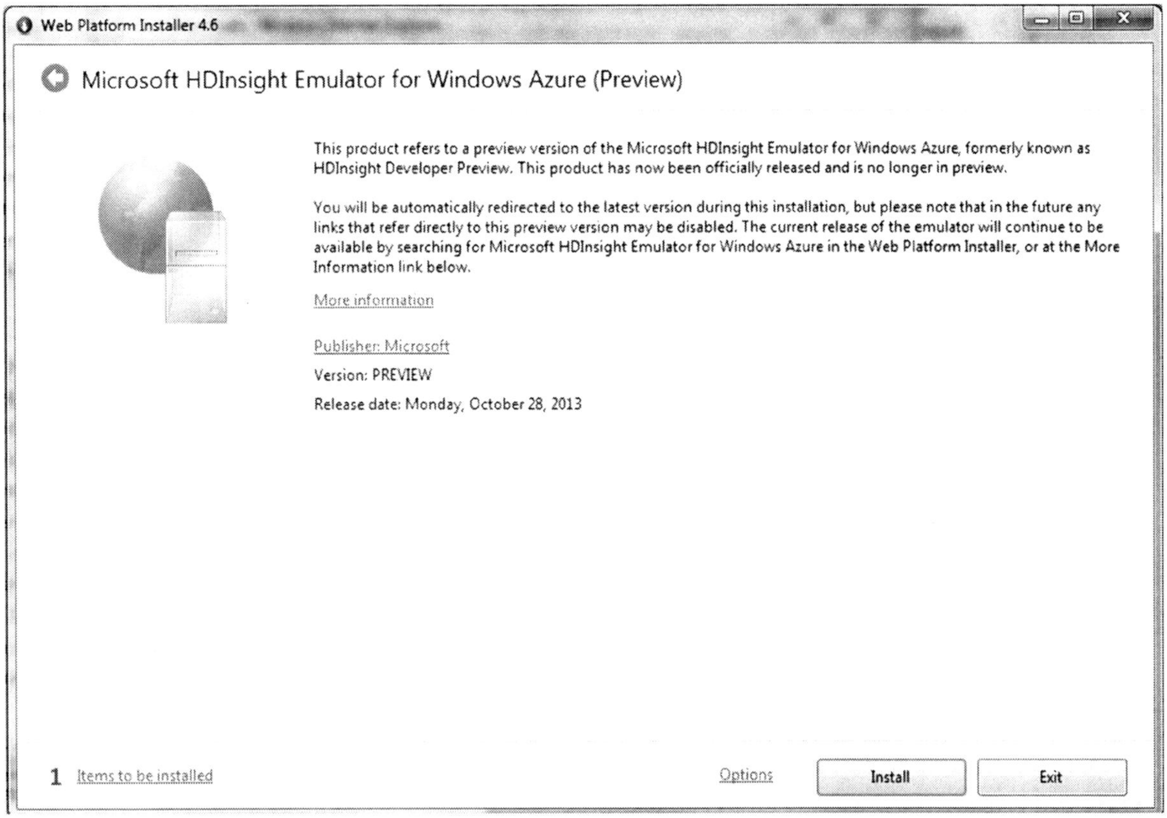

Figure 7-1. *Web PI*

Click on install, and accept the license terms to start the emulator installation. As stated earlier, it will download and install the Hortonworks Data Platform in your server, as shown in Figure 7-2.

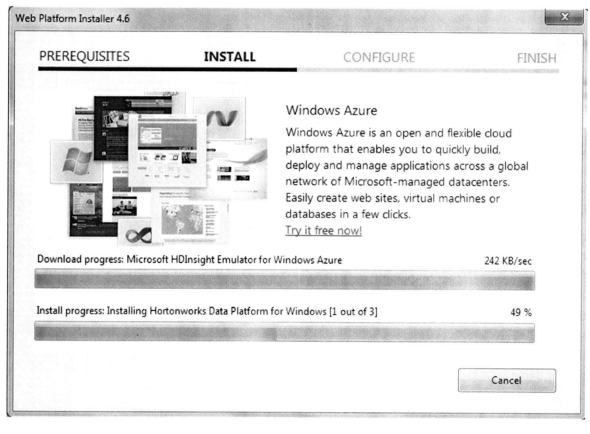

Figure 7-2. *Installing HDP*

■ **Note**   The HDInsight Emulator supports only the 64-bit flavor of the Windows OS family.

Verifying the Installation

Once the installation is complete, you can confirm if it is successful by verifying the presence of the Hadoop portal shortcuts on your desktop. Much like the Azure HDInsight name node, the emulator places the shortcuts to the Name Node status, the MapReduce status portals, and the Hadoop Command Line on the desktop, as shown in Figure 7-3.

Figure 7-3. *Hadoop portals*

You can also confirm the installation status from the *Control Panel* ➤ *Programs and Features* snap in. You are good if you find HDInsight Emulator and HDP in the list of installed programs as shown in Figure 7-4.

Hortonworks Data Platform 1.1 Developer   Hortonworks
Microsoft HDInsight Emulator for Windows Azure   Microsoft Corporation

Figure 7-4. *Programs and Features list*

If there is a problem with the installation, the first thing you should do is to go to the *Programs and Features* page in the Control Panel and check for these two items:

- Microsoft HDInsight Emulator for Windows Azure

- Hortonworks Data Platform 1.1 Developer

Uninstall these items, and repeat the installation procedure. The order of uninstall is important. You should uninstall the *Windows Azure HDInsight Emulator* first, and then the *Hortonworks Data Platform 1.1 Developer*.

The best approach to troubleshoot such installation/uninstallation issues is to enable MSI logging. You can follow the instructions in the following Knowledge Base article to set up MSI logging:

http://support.microsoft.com/2233000

After enabling logging repeat the action that failed and the log that's generated should point you in the right direction. If it turns out that uninstallation is failing due to missing setup files then you can probably try to get the missing files in place from another installation of the emulator.

Just to reiterate, the HDInsight emulator is a single-node deployment. So don't be surprised when you see that the number of live nodes in your cluster is *1* after you launch the Hadoop Name Node Status portal as shown in Figure 7-5.

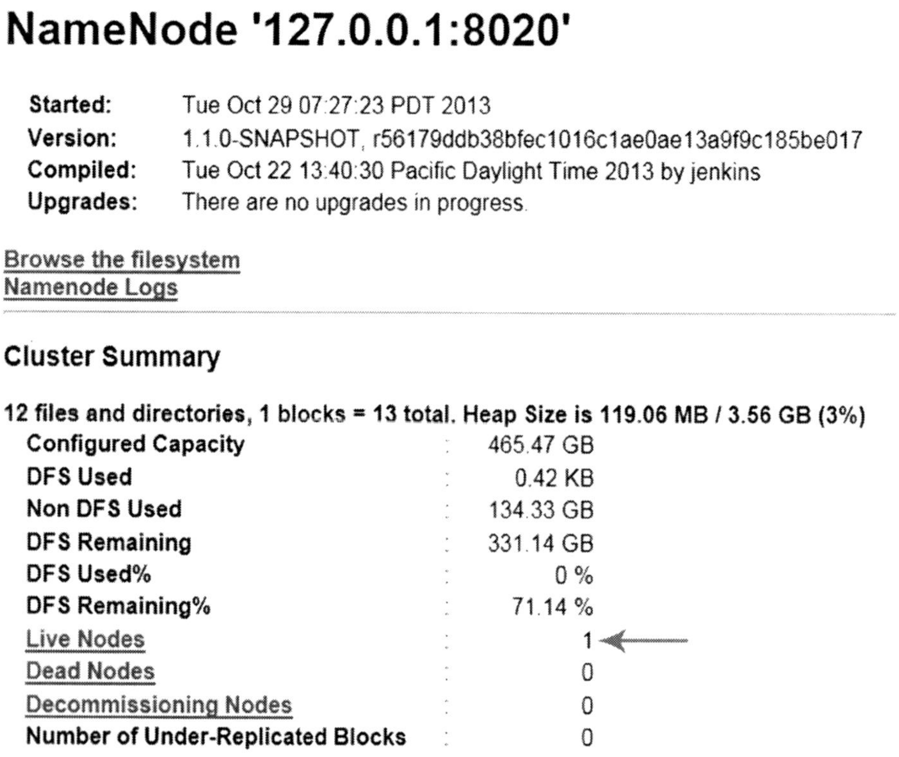

NameNode '127.0.0.1:8020'

Started: Tue Oct 29 07:27:23 PDT 2013
Version: 1.1.0-SNAPSHOT, r56179ddb38bfec1016c1ae0ae13a9f9c185be017
Compiled: Tue Oct 22 13:40:30 Pacific Daylight Time 2013 by jenkins
Upgrades: There are no upgrades in progress.

Browse the filesystem
Namenode Logs

Cluster Summary

12 files and directories, 1 blocks = 13 total. Heap Size is 119.06 MB / 3.56 GB (3%)

Configured Capacity	:	465.47 GB
DFS Used	:	0.42 KB
Non DFS Used	:	134.33 GB
DFS Remaining	:	331.14 GB
DFS Used%	:	0 %
DFS Remaining%	:	71.14 %
Live Nodes	:	1 ←
Dead Nodes	:	0
Decommissioning Nodes	:	0
Number of Under-Replicated Blocks	:	0

Figure 7-5. HDInsight Emulator Name Node portal

■ **Note** If you get errors launching the portal, make sure that the Apache Hadoop services are running through the Windows Services console (Start ➤ Run ➤ Services.msc).

The deployment of core Hadoop and the supporting projects is done in the C:\Hadoop\ directory by the emulator. Note that this path is slightly different: C:\apps\Dist\ directory in the case of the actual Azure HDInsight service. As of this writing, the Emulator ships version 1.6 of HDInsight service, which is HDP version 1.1. This is going to get updated periodically as and when the new versions of core Hadoop and HDP are tested and ported to the Windows platform. When you navigate to the C:\Hadoop directory, you should see a folder hierarchy similar to Figure 7-6.

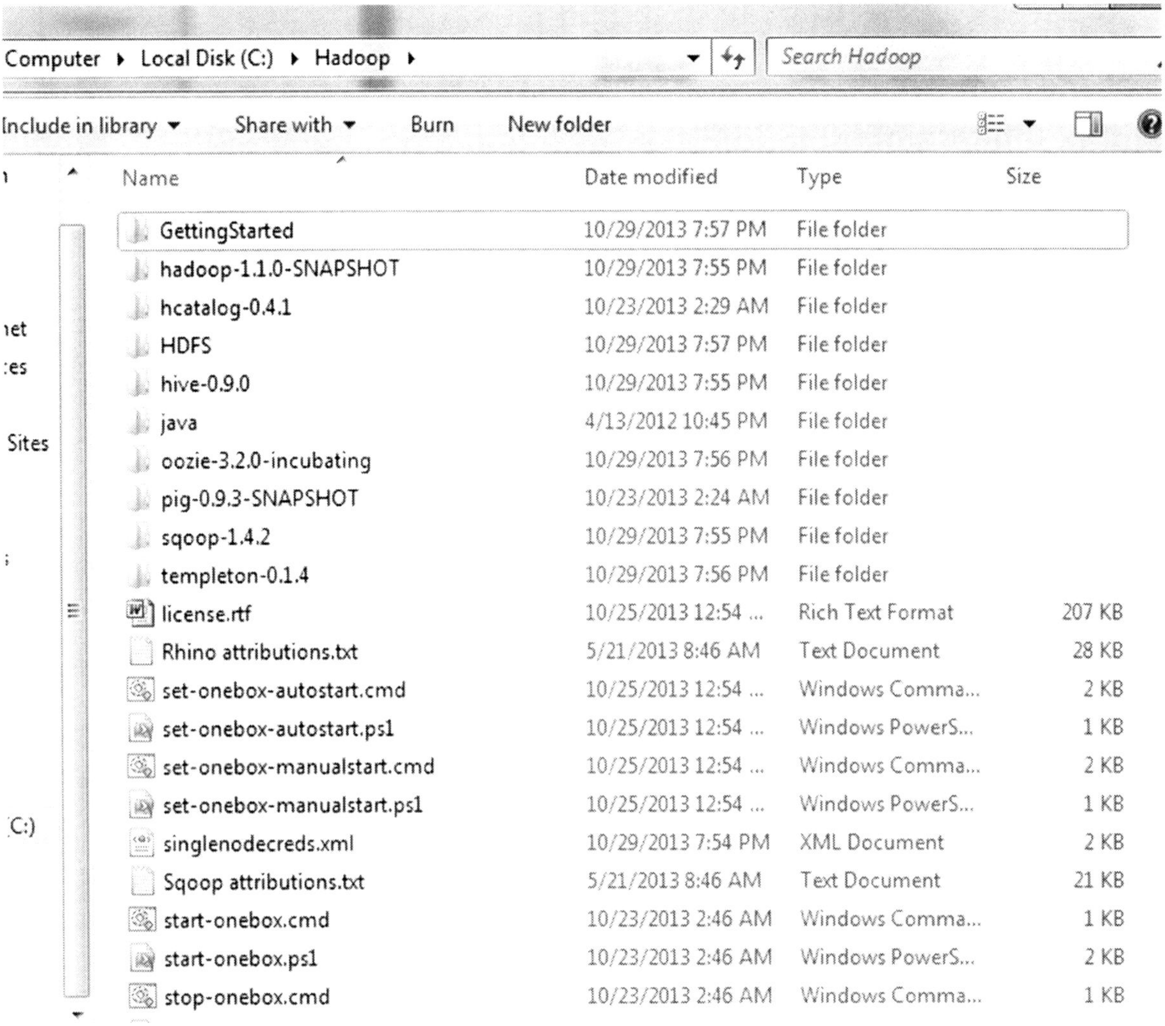
Computer ▸ Local Disk (C:) ▸ Hadoop ▸
Search Hadoop
Include in library ▾ Share with ▾ Burn New folder
Name Date modified Type Size
GettingStarted 10/29/2013 7:57 PM File folder
hadoop-1.1.0-SNAPSHOT 10/29/2013 7:55 PM File folder
hcatalog-0.4.1 10/23/2013 2:29 AM File folder
HDFS 10/29/2013 7:57 PM File folder
hive-0.9.0 10/29/2013 7:55 PM File folder
java 4/13/2012 10:45 PM File folder
oozie-3.2.0-incubating 10/29/2013 7:56 PM File folder
pig-0.9.3-SNAPSHOT 10/23/2013 2:24 AM File folder
sqoop-1.4.2 10/29/2013 7:55 PM File folder
templeton-0.1.4 10/29/2013 7:56 PM File folder
license.rtf 10/25/2013 12:54 ... Rich Text Format 207 KB
Rhino attributions.txt 5/21/2013 8:46 AM Text Document 28 KB
set-onebox-autostart.cmd 10/25/2013 12:54 ... Windows Comma... 2 KB
set-onebox-autostart.ps1 10/25/2013 12:54 ... Windows PowerS... 1 KB
set-onebox-manualstart.cmd 10/25/2013 12:54 ... Windows Comma... 2 KB
set-onebox-manualstart.ps1 10/25/2013 12:54 ... Windows PowerS... 1 KB
singlenodecreds.xml 10/29/2013 7:54 PM XML Document 2 KB
Sqoop attributions.txt 5/21/2013 8:46 AM Text Document 21 KB
start-onebox.cmd 10/23/2013 2:46 AM Windows Comma... 1 KB
start-onebox.ps1 10/23/2013 2:46 AM Windows PowerS... 2 KB
stop-onebox.cmd 10/23/2013 2:46 AM Windows Comma... 1 KB

Figure 7-6. *HDInsight Emulator installation directory*

Also, the logging infrastructure—along with the log files and paths—is exactly identical to what you see in the actual Azure service Name Node. Each of the project folders has its respective log directories that generate the service log files. For example, Figure 7-7 shows the Hadoop log files as generated by the Emulator installation.

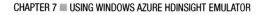

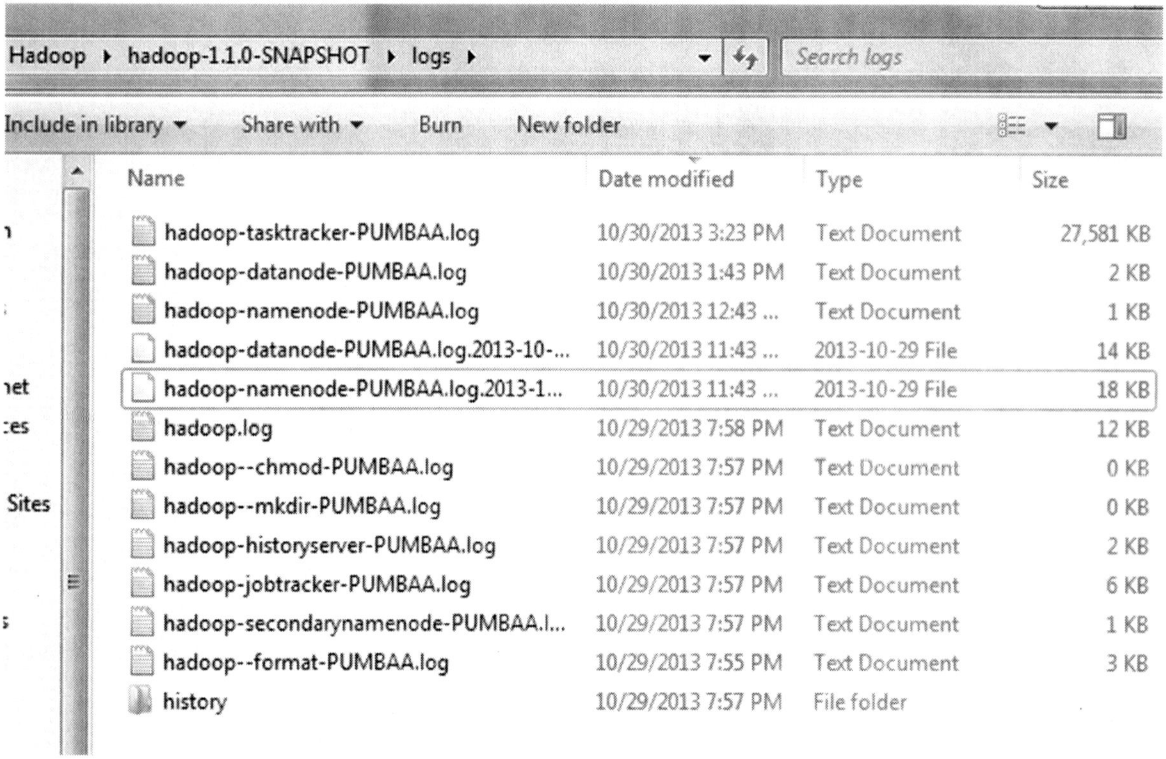

Figure 7-7. Hadoop log files

■ **Note** Details on HDInsight logging is explained in Chapter 11.

By default, the local emulator uses HDFS as its cluster storage. This can be changed by modifying the `core-site.xml` file in the `C:\Hadoop\hadoop-1.1.0-SNAPSHOT\conf` directory. You can add your Azure storage account key and container in the configuration file to point to Windows Azure Storage Blob (WASB). Listing 7-1 shows a sample entry in the `core-site.xml` file.

Listing 7-1. core-site.xml

```
<property>
    <name>fs.azure.account.key.democluster.blob.core.windows.net</name>
    <value>your_storage_account_key</value>
</property>
<property>
    <name>fs.default.name</name>
    <!-- cluster variant -->
    <value>wasb://democlustercontainer@democluster.blob.core.windows.net</value>
    <!--<value>hdfs://localhost:8020</value>-->
    <description>The name of the default file system.  Either the
```

```
literal string "local" or a host:port for NDFS.</description>
   <final>true</final>
 </property>
<property>
    <name>dfs.namenode.rpc-address</name>
    <value>hdfs://localhost:8020</value>
    <description>A base for other temporary directories.</description>
 </property>
```

■ **Note**   I have a storage account, democluster, and a default container, democlustercontainer. You may need to replace these values with your own.

There is also a way to emulate Azure blob storage in your local machine where you have installed the HDInsight emulator. You can use *Windows Azure Storage Emulator* to emulate the Windows Azure Storage blob (WASB) table and queue cloud services on your local machine. Doing so helps you in getting started with basic testing and evaluation locally without incurring the cost associated with cloud service. The Windows Azure Storage emulator comes as a part of Windows Azure SDK for .NET. This book, however, does not use the storage emulator, rather it uses actual WASB as HDInsight storage. Detailed instructions on configuring the storage emulator to be used from the HDInsight emulator can be found at:

http://www.windowsazure.com/en-us/documentation/articles/hdinsight-get-started-emulator/#blobstorage

The emulator also deploys the same set of Windows Services as the Azure service. You can open up the Windows Services console from Start ➤ Run ➤ Services.msc to start, stop, and set the startup type of the Apache Hadoop services as shown in Figure 7-8.

iption.	Name	Description	Status	Startup Ty
	Apache Hadoop derbyserver		Started	Automati
	Apache Hadoop historyserver		Started	Automati
	Apache Hadoop hiveserver		Started	Automati
	Apache Hadoop hiveserver2		Started	Automati
	Apache Hadoop hwi		Started	Automati
	Apache Hadoop jobtracker		Started	Automati
	Apache Hadoop metastore		Started	Automati
	Apache Hadoop namenode		Started	Automati
	Apache Hadoop oozieservice		Started	Automati
	Apache Hadoop secondarynamenode		Started	Automati
	Apache Hadoop tasktracker		Started	Automati
	Apache Hadoop templeton		Started	Automati

Figure 7-8. *The Apache Hadoop services*

There are, however, changes in the port numbers of the REST APIs that the emulator exposes. Logically enough, the security constraints are much less restrictive in the local emulator than with the Azure service, since the emulator resides in your local machine where you have more control. You have to be careful opening the respective ports if you wish to use the REST APIs to obtain status, version details, and so forth. Here is the list of the REST endpoints for the emulator, along with their port numbers:

- Oozie: `http://localhost:11000/oozie/v1/admin/status`.

- Templeton: `http://localhost:50111/templeton/v1/status`.

- ODBC: Use port 10001 in the DSN configuration or connection string.

- Cluster Name: Use `http://localhost:50111` as the cluster name wherever you require.

To start and stop the Hadoop services on the local emulator, you can use the `start-onebox.cmd` and `stop-onebox.cmd` command files from the `C:\Hadoop` directory. PowerShell versions for these files are available as well if you are a PowerShell fan, as shown in Figure 7-9.

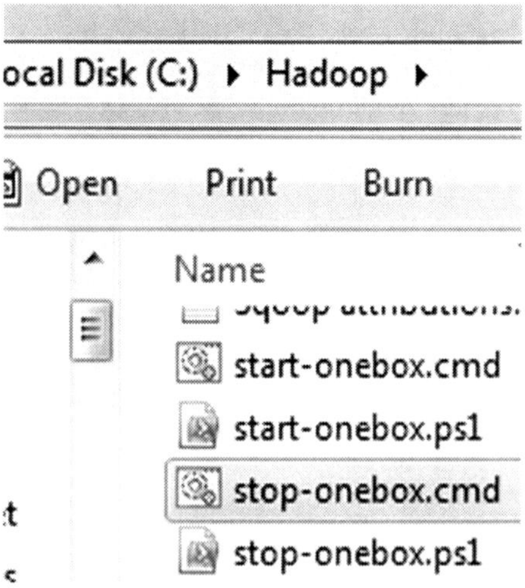

Figure 7-9. *Hadoop Service control files*

Once you start up the Hadoop services using the `start-onebox.cmd` file, you see output similar to Listing 7-2 in the console.

Listing 7-2. start-onebox.cmd

```
c:\Hadoop>start-onebox.cmd
Starting Hadoop Core services
Starting Hadoop services
Starting namenode
The Apache Hadoop namenode service is starting.
The Apache Hadoop namenode service was started successfully.

Starting datanode
The Apache Hadoop datanode service is starting.
The Apache Hadoop datanode service was started successfully.
```

```
Starting secondarynamenode
The Apache Hadoop secondarynamenode service is starting.
The Apache Hadoop secondarynamenode service was started successfully.

Starting jobtracker
The Apache Hadoop jobtracker service is starting.
The Apache Hadoop jobtracker service was started successfully.

Starting tasktracker
The Apache Hadoop tasktracker service is starting.
The Apache Hadoop tasktracker service was started successfully.

Starting historyserver
The Apache Hadoop historyserver service is starting.
The Apache Hadoop historyserver service was started successfully.

Starting Hive services
Starting hwi
The Apache Hadoop hwi service is starting.
The Apache Hadoop hwi service was started successfully.

Starting derbyserver
The Apache Hadoop derbyserver service is starting.
The Apache Hadoop derbyserver service was started successfully.

Starting metastore
The Apache Hadoop metastore service is starting.
The Apache Hadoop metastore service was started successfully.

Wait 10s for metastore db setup
Starting hiveserver
The Apache Hadoop hiveserver service is starting.
The Apache Hadoop hiveserver service was started successfully.

Starting hiveserver2
The Apache Hadoop hiveserver2 service is starting.
The Apache Hadoop hiveserver2 service was started successfully.

Starting Oozie service
Starting oozieservice...
Waiting for service to start
Oozie Service started successfully
The Apache Hadoop templeton service is starting.
The Apache Hadoop templeton service was started successfully.
```

Note Any service startup failures will also be displayed in the console. You may need to navigate to the respective log files to investigate further.

Using the Emulator

Working with the emulator is no different from using the Azure service except for a few nominal changes. Specifically, if you modified the core-site.xml file to point to your Windows Azure Blob Storage, there are very minimal changes to your Hadoop commands and MapReduce function calls. You can always use the Hadoop Command Line to execute your MapReduce jobs. For example, to list the contents of your storage Blob container, you can fire the ls command as shown in Listing 7-3.

Listing 7-3. Executing the ls command

```
hadoop fs -ls wasb://democlustercontainer@democluster.blob.core.windows.net/
```

You can also execute MapReduce jobs using the command line. Listing 7-4 shows you a sample job you can trigger from the Hadoop command prompt.

Listing 7-4. Using the Hadoop command line

```
hadoop jar "hadoop-examples.jar" "wordcount""/example/data/gutenberg/davinci.txt"
"/example/data/WordCountOutputEmulator"
```

■ **Note**   You need to have the hadoop-examples.jar file at the root of your Blob container to execute the job successfully.

As with the Azure service, the recommended way to submit and execute MapReduce jobs is through the .NET SDK or the PowerShell cmdlets. You can refer to Chapter 5 for such job submission and execution samples; there are very minor changes, like the cluster name, which is your local machine when you are using the emulator. Listing 7-5 shows a sample PowerShell script you can use for your MapReduce job submissions.

Listing 7-5. MapReduce PowerShell script

```
$creds = Get-Credential
$cluster = http://localhost:50111
$inputPath = "wasb://democlustercontainer@democluster.blob.core.windows.net/
example/data/gutenberg/davinci.txt"
$outputFolder = "wasb://democlustercontainer@democluster.blob.core.windows.net/
example/data/WordCountOutputEmulatorPS"
$jar = "wasb://democlustercontainer@democluster.blob.core.windows.net/hadoop-examples.jar"
$className = "wordcount"
$hdinsightJob = New-AzureHDInsightMapReduceJobDefinition -JarFile $jar -ClassName $className
-Arguments $inputPath, $outputPath

# Submit the MapReduce job
$wordCountJob = Start-AzureHDInsightJob -Cluster $cluster -JobDefinition
$hdinsightJob -Credential $creds
# Wait for the job to complete
Wait-AzureHDInsightJob -Job $wordCountJob -WaitTimeoutInSeconds 3600 -Credential $creds
```

■ **Note**   When prompted for credentials, provide `hadoop` as the user name and type in any text as the password. This is essentially a dummy credential prompt, which is needed to maintain compatibility with the Azure service from PowerShell scripts.

Future Directions

With hardware cost decreasing considerably over the years, organizations are leaning toward appliance-based, data-processing engines. An appliance is a combination of hardware units and built-in software programs suitable for a specific kind of workload. Though Microsoft has no plans to offer a multinode HDInsight solution for use on premises, it does offer an appliance-based multiunit, and massively parallel processing (MPP) device, called the Parallel Data Warehouse (PDW). Microsoft PDW gives you performance and scalability for data warehousing with the plug and play simplicity of an appliance. Some nodes in the appliance can run SQL PDW, and some nodes can run Hadoop (called a *Hadoop Region*). A new data-processing technology called *Polybase* has been introduced, which is designed to be the simplest way to combine nonrelational data and traditional relational data for your analysis. It acts as a bridge to allow SQL PDW to send queries to Hadoop and fetch data results. The nice thing is that users can send regular SQL queries to PDW, and Hadoop can run them and fetch data from unstructured files. To learn more about PDW and Polybase, see the following MSDN page:

`http://www.microsoft.com/en-us/sqlserver/solutions-technologies/data-warehousing/polybase.aspx`

The Open Source Apache Hadoop project is going through a lot of changes as well. In the near future, Hadoop version 2.0 will be widely available. Hadoop 2.0 introduces a new concept called *Yet Another Resource Negotiator (YARN)* on top of traditional MapReduce. This is also known as MapReduce 2.0 or MRv2. With HDInsight internally using Hadoop, it is highly likely that the Azure Service and the Emulator will be upgraded to Hadoop 2.0 as well in due course. The underlying architecture, however, will be the same in terms job submissions and end-user interactions; hence, the impact of this change to the readers and users will be minimal.

Summary

The HDInsight offering is essentially a cloud service from Microsoft. Since even evaluating the Windows Azure HDInsight Service involves some cost, an emulator is available as a single-node box product on your Windows Server systems, which you can use as your playground to test and evaluate the technology. The Windows Azure HDInsight Emulator uses the same software bits as the Azure Service and supports the exact same set of functionality. It is designed to be scalable and perform massive parallel processing, so you can test your Big Data solution on the emulator. Once you are satisfied, you can deploy your actual solution to production in Azure and take advantage of multinode Hadoop clusters on Windows Azure. For on-premises use, Microsoft is offering its Parallel Data Warehouse (PDW) technology, which is an appliance-based multinode HDInsight cluster, while the emulator will continue to be single node and serve as a test bed.

■ ■ ▪

## CHAPTER 8

Accessing HDInsight over Hive and ODBC

If you are a SQL developer and want to cross-pollinate your existing SQL skills in the world of Big Data, Hive is probably the best place for you. This section of the book will enable you to be the Queen Bee of your Hadoop world with Hive and gain business intelligence (BI) insights with Hive Query Language (HQL) filters and joins of Hadoop Distributed File System (HDFS) datasets.

Hive provides a schema to the underlying HDFS data and a SQL-like query language to access that data. Simba, in collaboration with Microsoft, provides an ODBC driver that is the supported and recommended interface for connecting to HDInsight. It can enable client applications to connect and consume Hive data that resides on top of your HDFS (WASB, in case of HDInsight). The driver is available for a free download at:

```
http://www.microsoft.com/en-us/download/details.aspx?id=40886
```

The preceding link has both the 32-bit and 64-bit Hive ODBC drivers available for download. You should download the appropriate version of the driver for your operating system and the application that will consume the driver, and be sure to match the bitness. For example, if you want to consume the driver from the 32-bit version of Excel, you will need to install the 32-bit Hive ODBC driver.

This chapter shows you how to create a basic schema structure in Hive, load data into that schema, and access the data using the ODBC driver from a client application.

Hive: The Hadoop Data Warehouse

Hive is a framework that sits on top of core Hadoop. It acts as a data-warehousing system on top of HDFS and provides easy query mechanisms to the underlying HDFS data. By revisiting the Hadoop Ecosystem diagram in Chapter 1, you can see that Hive sits right on top of Hadoop core, as shown in Figure 8-1.

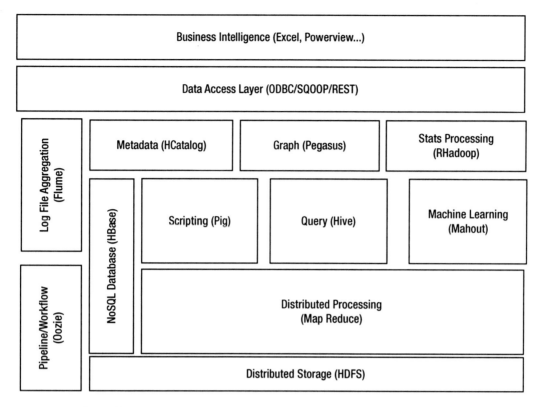

Figure 8-1. *The Hadoop ecosystem*

Programming MapReduce jobs can be tedious, and they require their own development, testing, and maintenance investments. Hive lets you democratize access to Big Data using familiar tools such as Excel and a SQL-like language without having to write complex MapReduce jobs. Hive queries are broken down into MapReduce jobs under the hood, and they remain a complete abstraction to the user. The simplicity and SQL-ness of Hive queries has made Hive a popular and preferred choice for users. That is particularly so for users with traditional SQL skills, because the ramp-up time is so much less than what is required to learn how to program MapReduce jobs directly.

Figure 8-2 gives an overview of the Hive architecture.

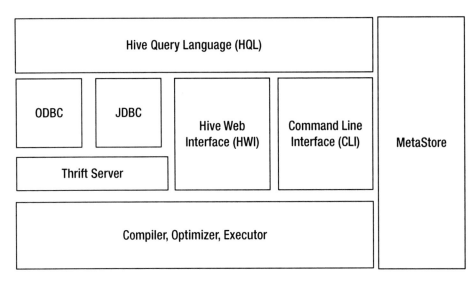

Figure 8-2. *Hive architecture*

In effect, Hive enables you to create an interface layer over MapReduce that can be used in a similar fashion to a traditional relational database. This layer enables business users to use familiar tools like Excel and SQL Server Reporting Services to consume data from HDInsight as they would from a database system such as SQL Server, remotely through an ODBC connection. The rest of this chapter walks you through different Hive operations and using the Hive ODBC driver to consume the data.

Working with Hive

Hive uses tables to impose schema on data and provides a query interface for client applications. The key difference between Hive tables and those in traditional database systems, such as SQL Server, is that Hive adopts a *schema on read* approach. This approach enables you to be flexible about the specific columns and data types that you want to project on top of your data. You can create multiple tables with different schema from the same underlying data, depending on how you want to use that data. The most important point to take away from this approach is that the table is simply a metadata schema that is imposed on data in underlying files.

Creating Hive Tables

You create tables by using the HiveQL CREATE TABLE statement, which in its simplest form looks similar to the analogous statement in Transact-SQL.

One thing to note about Hive tables is that you can create two types of tables: *External* and *Internal*. If you do not specify a table type, a table is created as *Internal*.

Be careful! An internal table tells Hive to manage the data by itself. If you drop the table, by default the data is also dropped and cannot be recovered. If you want to manage the data and data locations, if your data is used outside Hive, or if you need to retain the data, create an external table. The syntax is pretty much similar, requiring just the addition of the EXTERNAL keyword.

You can use the PARTITIONED BY clause to create a subfolder for each distinct value in a specified column (for example, to store a file of daily data for each date in a separate folder). Partitioning can improve query performance because HDInsight will scan only relevant partitions in a filtered query.

You can use the SKEWED BY clause to create separate files for each row where a specified column value is in a list of specified values. Rows with values not listed are stored in a single other file.

You can use the CLUSTERED BY clause to distribute data across a specified number of subfolders (described as *buckets*) based on the values of specified columns using a hashing algorithm.

There are a few of ways to execute Hive queries against your HDInsight cluster:

- Using the Hadoop Command Line

- Using .NET SDK

- Using Windows Azure PowerShell

In this chapter, we use Windows Azure PowerShell to create, populate, and query Hive tables. The Hive tables are based on some demo stock data of different companies as specified here:

- Apple

- Facebook

- Google

- MSFT

- IBM

- Oracle

Let's first load the input files to the WASB that our democluster is using by executing the following PowerShell script in Listing 8-1. The input files used in this book are just a subset of the stock market dataset available for free at www.infochimps.com and is provided separately.

Listing 8-1. Uploading files to WASB

```
$subscriptionName = "<YourSubscriptionname>"
$storageAccountName = "democluster"
$containerName = "democlustercontainer"
#This path may vary depending on where you place the source .csv files.
$fileName ="D:\HDIDemoLab\TableFacebook.csv"
$blobName = "Tablefacebook.csv"
# Get the storage account key
Select-AzureSubscription $subscriptionName
$storageaccountkey = get-azurestoragekey $storageAccountName | %{$_.Primary}
# Create the storage context object
$destContext = New-AzureStorageContext –StorageAccountName
         $storageAccountName -StorageAccountKey $storageaccountkey
# Copy the file from local workstation to the Blob container
Set-AzureStorageBlobContent -File $fileName -Container $containerName
         -Blob $blobName -context $destContext
```

■ **Note** Repeat these steps with other .csv files in the folder by changing the $filename variable and $blobname variables and rerun Set-AzureStorageBlobContent.

Once the files are uploaded, log on to the name node using remote desktop and execute the command `hadoop fs -ls /` in the Hadoop Command Line. This should list all the files you just uploaded, as shown in Listing 8-2.

Listing 8-2. Listing the uploaded files

```
c:\apps\dist\hadoop-1.2.0.1.3.1.0-06>hadoop fs -ls /
Found 10 items
-rwxrwxrwx   1        15967 2013-11-24 06:43 /TableFacebook.csv
-rwxrwxrwx   1       130005 2013-11-24 06:42 /TableGoogle.csv
-rwxrwxrwx   1       683433 2013-11-24 06:42 /TableIBM.csv
-rwxrwxrwx   1       370361 2013-11-24 06:43 /TableMSFT.csv
-rwxrwxrwx   1       341292 2013-11-24 06:42 /TableOracle.csv
-rwxrwxrwx   1       341292 2013-11-24 06:43 /TableApple.csv
```

You can also use the Azure portal to navigate to the storage account container, and you should be able to see the files uploaded, as shown in Figure 8-3.

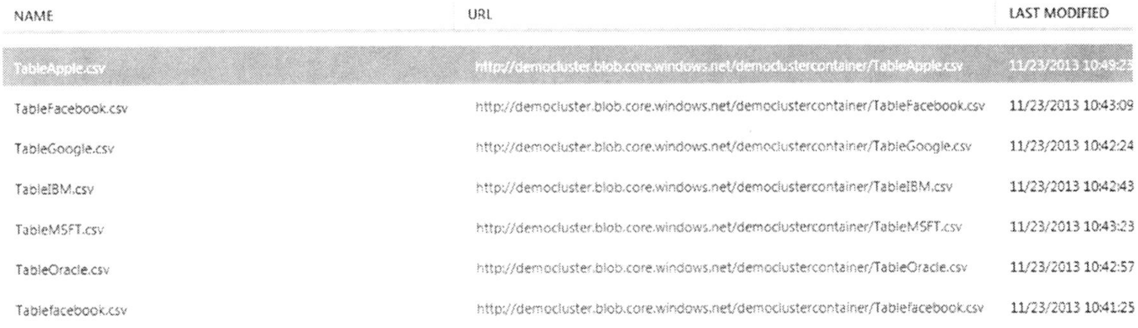

NAME	URL	LAST MODIFIED
TableApple.csv	http://democluster.blob.core.windows.net/democlustercontainer/TableApple.csv	11/23/2013 10:49:23
TableFacebook.csv	http://democluster.blob.core.windows.net/democlustercontainer/TableFacebook.csv	11/23/2013 10:43:09
TableGoogle.csv	http://democluster.blob.core.windows.net/democlustercontainer/TableGoogle.csv	11/23/2013 10:42:24
TableIBM.csv	http://democluster.blob.core.windows.net/democlustercontainer/TableIBM.csv	11/23/2013 10:42:43
TableMSFT.csv	http://democluster.blob.core.windows.net/democlustercontainer/TableMSFT.csv	11/23/2013 10:43:23
TableOracle.csv	http://democluster.blob.core.windows.net/democlustercontainer/TableOracle.csv	11/23/2013 10:42:57
Tablefacebook.csv	http://democluster.blob.core.windows.net/democlustercontainer/Tablefacebook.csv	11/23/2013 10:41:25

Figure 8-3. *The democlustercontainer*

Note that the files are uploaded to the root directory. To make it more structured, we will copy the stock data files into the `StockData` folder. With Remote Desktop, open the Hadoop Command Line and execute commands shown in Listing 8-3.

Listing 8-3. Copying the data files to `StockData` folder

```
hdfs fs -cp /TableApple.csv /debarchan/StockData/tableApple.csv
hdfs fs -cp /TableFacebook.csv /debarchan/StockData/tableFacebook.csv
hdfs fs -cp /TableGoogle.csv /debarchan/StockData/tableGoogle.csv
hdfs fs -cp /TableIBM.csv /debarchan/StockData/tableIBM.csv
hdfs fs -cp /TableMSFT.csv /debarchan/StockData/tableMSFT.csv
hdfs fs -cp /TableOracle.csv /debarchan/StockData/tableOracle.csv
```

■ **Note** The file and folder names are case sensitive. Also, you will need to replace the user name value with the one you configured for Remote Desktop access.

This will copy all the .csv files under the /debarchan/StockData folder.

Once the source files are staged in your WASB, you need to define the Hive schema that will be a placeholder for your Hive tables when you actually load data into it. Note that to run the PowerShell commands, you have to download and install Windows Azure HDInsight PowerShell as described in Chapter 4. The HDInsight PowerShell modules are integrated with Windows Azure Powershell version 0.7.2 and can be downloaded from http://www.windowsazure.com/en-us/documentation/articles/hdinsight-install-configure-powershell/.

Execute the command in Listing 8-4 to create the Hive table.

Listing 8-4. Creating the Hive table stock_analysis

```
$subscriptionName = "YourSubscriptionName"
$storageAccountName = "democluster"
$containerName = "democlustercontainer"
$clustername = "democluster"

$querystring = "create external table stock_analysis
        (stock_symbol string,stock_Date string,
        stock_price_open double,stock_price_high  double,
        stock_price_low double,
        stock_price_close double,
        stock_volume int,
        stock_price_adj_close double)
        partitioned by (exchange string)
        row format delimited
        fields terminated by ','
        LOCATION 'wasb://democlustercontainer@democluster.blob.core.windows.net/debarchan/StockData';"
```

▪ **Note** You may need to wrap each of the commands in single line, depending on the PowerShell editor you use. Otherwise, you may encounter syntactical errors while running the script.

```
$HiveJobDefinition = New-AzureHDInsightHiveJobDefinition -Query $querystring
$HiveJob = Start-AzureHDInsightJob -Subscription $subscriptionname
        -Cluster $clustername -JobDefinition $HiveJobDefinition

$HiveJob | Wait-AzureHDInsightJob -Subscription $subscriptionname
        -WaitTimeoutInSeconds 3600

Get-AzureHDInsightJobOutput -Cluster $clustername
 -Subscription $subscriptionname
    -JobId $HiveJob.JobId -StandardError
```

Once the job execution is complete, you should see output similar to the following:

```
StatusDirectory : 2b391c76-2d33-42c4-a116-d967eb11c115
ExitCode        : 0
Name            : Hive: create external table
Query           : create external table stock_analysis
            (stock_symbol string,stock_Date string,
            stock_price_open double,stock_price_high double,
```

```
        stock_price_low double,stock_price_close double,
        stock_volume int,stock_price_adj_close double)
        partitioned by (exchange string)
        row format delimited
        fields terminated by ','
        LOCATION wasb://democlustercontainer@democluster.blob.core.windows.net/debarchan/StockData';


State         : Completed
SubmissionTime : 11/24/2013 7:08:25 AM
Cluster       : democluster
PercentComplete :
JobId         : job_201311240635_0002


Logging initialized using configuration in file:/C:/apps/dist/hive-0.11.0.1.3.1.0-06/
conf/hive-log4j.properties
OK
Time taken: 22.438 seconds
```

You can verify the structure of the schema you just created using the script in Listing 8-5.

Listing 8-5. Verifying the Hive schema

```
$subscriptionName = "YourSubscriptionName"
$clustername = "democluster"
Select-AzureSubscription -SubscriptionName $subscriptionName
Use-AzureHDInsightCluster $clusterName
    -Subscription (Get-AzureSubscription -Current).SubscriptionId


$querystring = "DESCRIBE stock_analysis;"
Invoke-Hive -Query $querystring
```

This should display the structure of the stock_analysis table as shown here:

```
Successfully connected to cluster democluster
Submitting Hive query..
Started Hive query with jobDetails Id : job_201311240635_0004
Hive query completed Successfully


stock_symbol            string      None
stock_date              string      None
stock_price_open        double      None
stock_price_high        double      None
stock_price_low         double      None
stock_price_close       double      None
stock_volume            int         None
stock_price_adj_close   double      None
exchange                string      None

# Partition Information
# col_name               data_type  comment
exchange                string      None
```

Now that you have the Hive schema ready, you can start loading the stock data in your stock_analysis table.

Loading Data

You can feed data to your Hive tables by simply copying data files into the appropriate folders. A table's definition is purely a metadata schema that is applied to the data files in the folders when they are queried. This makes it easy to define tables in Hive for data that is generated by other processes and deposited in the appropriate folders when ready.

Additionally, you can use the HiveQL LOAD statement to load data from an existing file into a Hive table. This statement moves the file from its current location to the folder associated with the table. LOAD does not do any transformation while loading data into tables. LOAD operations are currently pure copy/move operations that move data files into locations corresponding to Hive tables. This is useful when you need to create a table from the results of a MapReduce job or Pig script that generates an output file alongside log and status files. The technique enables you to easily add the output data to a table without having to deal with additional files you do not want to include in the table.

For example, Listing 8-6 shows how to load data into the analysis_stock table created earlier. You can execute the following PowerShell script, which will load data from TableMSFT.csv.

Listing 8-6. Loading data to a Hive table

```
$subscriptionName = "YourSubscriptionName"
$storageAccountName = "democluster"
$containerName = "democlustercontainer"
$clustername = "democluster"

$querystring = "load data inpath   'wasb://democlustercontainer@democluster.blob.core.windows.net/
                            debarchan/StockData/tableMSFT.csv'
       into table stock_analysis partition(exchange ='NASDAQ');"

$HiveJobDefinition = New-AzureHDInsightHiveJobDefinition
              -Query $querystring
$HiveJob = Start-AzureHDInsightJob -Subscription
       $subscriptionname -Cluster $clustername
              -JobDefinition $HiveJobDefinition

$HiveJob | Wait-AzureHDInsightJob -Subscription $subscriptionname
       -WaitTimeoutInSeconds 3600
Get-AzureHDInsightJobOutput -Cluster $clustername
       -Subscription $subscriptionname
              -JobId $HiveJob.JobId -StandardError
```

■ **Note**   You may need to wrap up each of the commands in single line to avoid syntax errors depending on the PowerShell editor you use.

You should see output similar to the following once the job completes:

```
StatusDirectory : 0b2e0a0b-e89b-4f57-9898-3076c10fddc3
ExitCode        : 0
Name            : Hive: load data inpath 'wa
Query           : load data inpath 'wasb://democlustercontainer@democluster.blob.core.windows.net/
                            debarchan/StockData/tableMSFT.csv' into table stock_analysis
                  partition(exchange ='NASDAQ');
```

```
State          : Completed
SubmissionTime : 11/24/2013 7:35:18 AM
Cluster        : democluster
PercentComplete :
JobId          : job_201311240635_0006

Logging initialized using configuration in file:/C:/apps/dist/hive-0.11.0.1.3.1.0-06/
conf/hive-log4j.properties
Loading data to table default.stock_analysis partition (exchange=NASDAQ)
OK
Time taken: 44.327 seconds
```

Repeat the preceding steps for all the .csv files you have to load into the table. Note that you need to replace only the .csv file names in $querystring and make sure you load the data into the respective partitions of the Hive table.

Listing 8-7 gives you all the LOAD commands for each of the .csv files.

Listing 8-7. The LOAD commands

```
$querystring = "load data inpath 'wasb://democlustercontainer@democluster.blob.core.windows.net/
                          debarchan/StockData/tableFacebook.csv'
      into table stock_analysis partition(exchange ='NASDAQ');"
$querystring = "load data inpath 'wasb://democlustercontainer@democluster.blob.core.windows.net/
                          debarchan/StockData/tableApple.csv'
      into table stock_analysis partition(exchange ='NASDAQ');"

$querystring = "load data inpath 'wasb://democlustercontainer@democluster.blob.core.windows.net/
                          debarchan/StockData/tableGoogle.csv'
      into table stock_analysis partition(exchange ='NASDAQ');"

$querystring = "load data inpath 'wasb://democlustercontainer@democluster.blob.core.windows.net/
                          debarchan/StockData/tableIBM.csv'
      into table stock_analysis partition(exchange ='NYSE');"

$querystring = "load data inpath 'wasb://democlustercontainer@democluster.blob.core.windows.net/
                          debarchan/StockData/tableOracle.csv'
      into table stock_analysis partition(exchange ='NYSE');"
```

Querying Tables with HiveQL

After you create tables and load data files into the appropriate locations, you can start to query the data by executing HiveQL SELECT statements against the tables. As with all data processing on HDInsight, HiveQL queries are implicitly executed as MapReduce jobs to generate the required results. HiveQL SELECT statements are similar to SQL queries, and they support common operations such as JOIN, UNION, and GROUP BY.

For example, you can use the code in Listing 8-8 to filter by stock_symbol column and also to return 10 rows for sampling, because you don't know how many rows you may have.

Listing 8-8. Querying data from a Hive table

```
$subscriptionName = "YourSubscriptionName"
$clustername = "democluster"
Select-AzureSubscription -SubscriptionName $subscriptionName
Use-AzureHDInsightCluster $clusterName
    -Subscription (Get-AzureSubscription -Current).SubscriptionId

$querystring  = "select * from stock_analysis
        where stock_symbol LIKE 'MSFT' LIMIT 10;"
Invoke-Hive -Query $querystring
```

You should see output similar to the following once the job execution completes:

```
Successfully connected to cluster democluster
Submitting Hive query..
Started Hive query with jobDetails Id : job_201311240635_0014
Hive query completed Successfully
```

MSFT	2/8/2013	31.69	31.9	31.57	31.89	29121500	31.89	NASDAQ
MSFT	1/8/2013	32.06	32.09	31.6	31.67	42328400	31.67	NASDAQ
MSFT	31/07/2013	31.97	32.05	31.71	31.84	43898400	31.84	NASDAQ
MSFT	30/07/2013	31.78	32.12	31.55	31.85	45799500	31.85	NASDAQ
MSFT	29/07/2013	31.47	31.6	31.4	31.54	28870700	31.54	NASDAQ
MSFT	26/07/2013	31.26	31.62	31.21	31.62	38633600	31.62	NASDAQ
MSFT	25/07/2013	31.62	31.65	31.25	31.39	63213000	31.39	NASDAQ
MSFT	24/07/2013	32.04	32.19	31.89	31.96	52803100	31.96	NASDAQ
MSFT	23/07/2013	31.91	32.04	31.71	31.82	65810400	31.82	NASDAQ
MSFT	22/07/2013	31.7	32.01	31.6	32.01	79040700	32.01	NASDAQ

It is very important to note that Hive queries use minimal caching, statistics, or optimizer tricks. They generally read the entire data set on each execution, and thus are more suitable for batch processing than for online work. One of the strongest recommendations I have for you while you are querying Hive is to write SELECT * instead of listing specific column names. Fetching a selective list of columns like in Listing 8-9 is a best practice when the source is a classic database management system like SQL Server database, but the story is completely different with Hive.

Listing 8-9. Selecting a partial list of columns

```
SELECT stock_symbol, stock_volume
FROM stock_analysis;
```

The general principle of HIVE is to expose Hadoop MapReduce functionality through an SQL-like language. Thus, when you issue a command like that in Listing 8-9 a MapReduce job will be triggered to remove any columns from the Hive table data set that aren't being specified in the query, and to send back only the columns stock_symbol and stock_volume.

On the other hand, the HiveQL in Listing 8-10 does not require any MapReduce job to return its results, because there is no need to eliminate columns. Hence, there is less processing in the background.

Listing 8-10. Selecting all columns

```
SELECT * FROM stock_analysis;
```

■ **Note**   In cases where selecting only a few columns reduces a lot of the data to transfer, it may still be interesting to select only a few columns.

In addition to common SQL semantics, HiveQL supports the inclusion of custom MapReduce scripts embedded in a query through the MAP and REDUCE clauses, as well as custom User Defined Functions (UDFs) implemented in Java. This extensibility enables you to use HiveQL to perform complex transforms to data as it is queried.

For a complete reference on Hive data types and HQL, see the Apache Hive language manual site:

```
https://cwiki.apache.org/confluence/display/Hive/Home
```

Hive Storage

Hive stores all its metadata in its storage, called a *Hive MetaStore*. Traditional Hive uses its native Derby database by default, but Hive can also be configured to use MySQL as its MetaStore. With HDInsight, this capability extends and the Hive MetaStore can be configured to be SQL Server as well as SQL Azure. You can modify the Hive configuration file hive-site.xml found under the conf folder in the Hive installation directory to customize your MetaStore. You can also customize the Hive MetaStore while deploying your HDInsight cluster through the *CUSTOM CREATE* wizard, which is explained in Chapter 3.

The Hive ODBC Driver

One of the main advantages of Hive is that it provides a querying experience that is similar to that of a relational database, which is a familiar experience for many business users. Additionally, the ODBC driver for Hive enables users to connect to HDInsight and execute HiveQL queries from familiar tools like Excel, SQL Server Integration Services (SSIS), PowerView, and others. Essentially, the driver allows all ODBC-compliant clients to consume HDInsight data through familiar ODBC Data Source Names (DSNs), thus exposing HDInsight to a wide range of client applications.

Installing the Driver

The driver comes in two flavors: 64 bit and 32 bit. Be sure to *install both the 32-bit and 64-bit versions of the driver*—you'll need to install them separately. If you install only the 64-bit driver, you'll get errors in your 32-bit applications—for example, Visual Studio when trying to configure your connections. The driver can be downloaded and installed from the following site:

```
http://www.microsoft.com/en-us/download/details.aspx?id=40886
```

Once the installation of the driver is complete, you can confirm the installation status by checking if you have the *Microsoft Hive ODBC Driver* present in the ODBC Data Source Administrator's list of drivers, as shown in Figure 8-4.

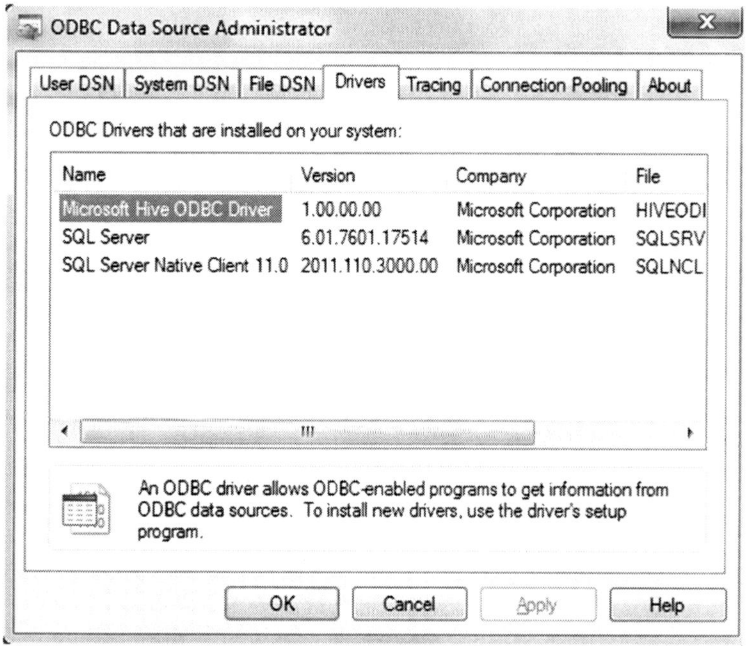

Figure 8-4. *ODBC Data Source Administrator*

■ **Note**   There are two versions of the ODBC Data Source Administrator UI: one for 32-bit (*%windir%\SysWOW64\ odbcad32.exe*) and one for 64-bit (*%windir%\System32\odbcad32.exe*). You'll likely want to create both 32-bit and 64-bit DSNs—just make sure that the same name is used for both versions. At a minimum, you'll need to register a 32-bit DSN to use when creating your SSIS package in the designer in Chapter 10.

The presence of the Microsoft Hive ODBC driver under the list of available ODBC drivers ensures that it has been installed successfully.

Testing the Driver

Once the driver is installed successfully, the next step is to ensure that you can make a connection to Hive using the driver. First, create a System DSN. In ODBC Data Sources Administrator, go to the *System DSN* tab and click on the *Add* button as shown in Figure 8-5.

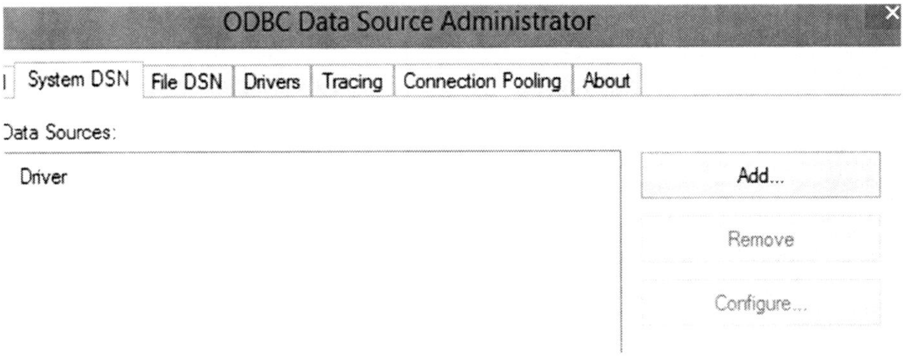

Figure 8-5. *Add System DSN*

Choose the *Microsoft Hive ODBC Driver* driver in the next screen of the *Create New Data Source* wizard, as shown in Figure 8-6.

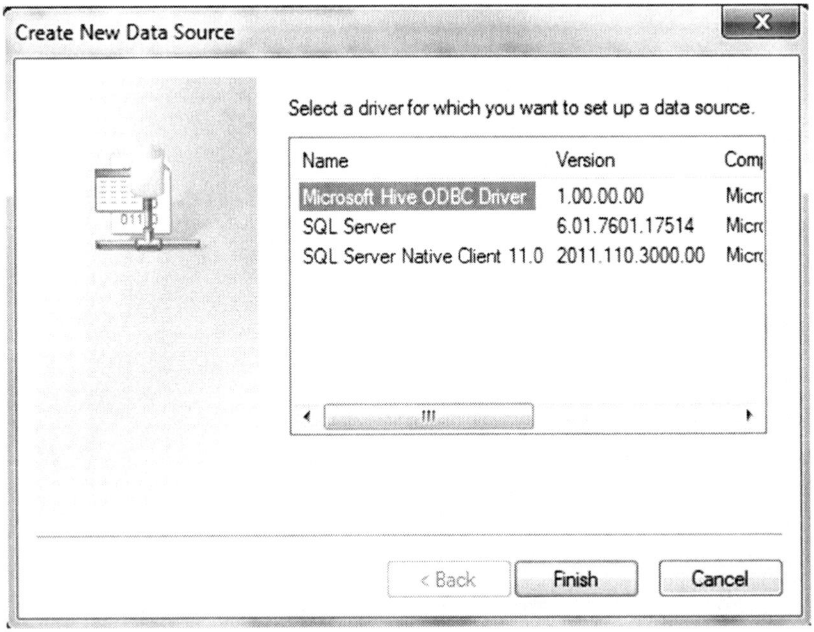

Figure 8-6. *Selecting the Microsoft Hive ODBC Driver*

After clicking *Finish*, you are presented with the final *Microsoft Hive ODBC Driver DSN Setup* screen, where you'll need to provide the following:

- Host: This is the full domain name of your HDInsight cluster (*democluster.azurehdinsight.net*)

- Port: *443* is the default

- Database: *default*

- Hive Server Type: *Hive Server 2*

- Authentication Mechanism: Select *Windows Azure HDInsight Service*

- User Name & Password: This will be the user name and password you used while creating your cluster

Enter the HDInsight cluster details as well as the credentials used to connect to the cluster. In this sample, I am using my HDInsight cluster deployed in Windows Azure, as shown in Figure 8-7.

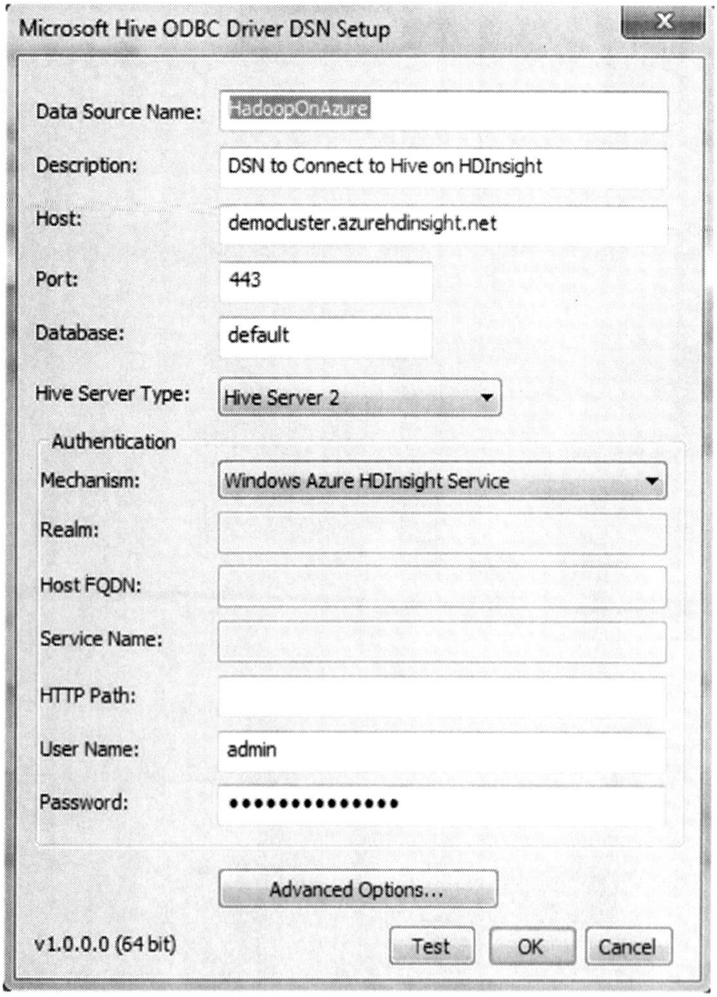

Figure 8-7. *Finishing the configuration*

Click on the *Test* button to make sure that a connection could be established successfully, as shown in Figure 8-8.

Figure 8-8. *Testing a connection*

There are a few settings of interest on the *Advanced Options* page of the DSN Setup screen. The most important one is the *Default string column length* value. By default, this will be set to 65536, which is larger than the maximum string length of many client applications—for example, SSIS—which may have negative performance implications. If you know that your data values will be less than the maximum characters in length supported by your client application, I recommend lowering this value to 4000 (or less).

The other options you can control through the *Advanced Options* page are *Rows fetched per block*, *Binary column length*, *Decimal column scale*, usage of *Secure Sockets Layer (SSL)* certificates, and so on, as shown in Figure 8-9.

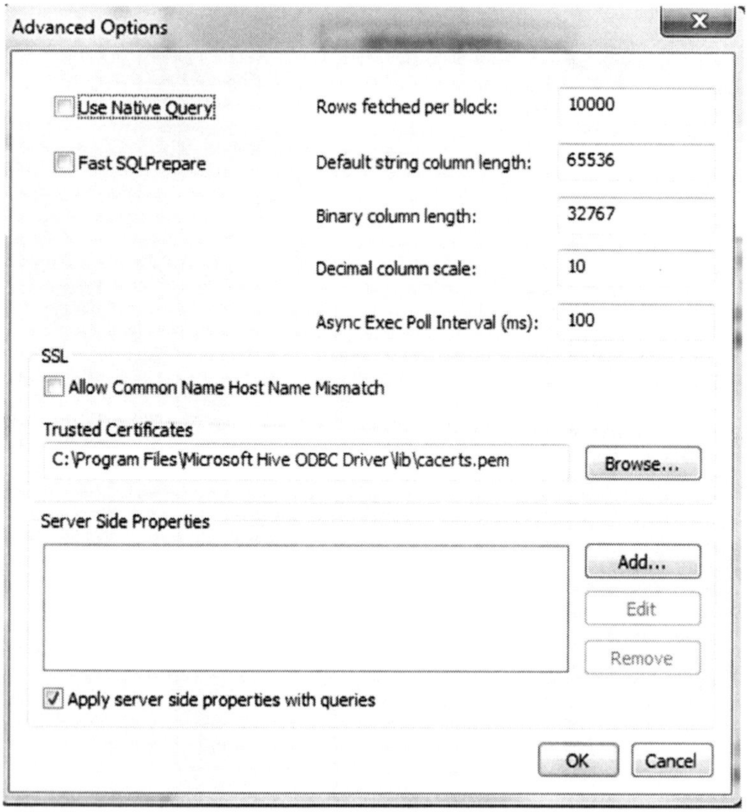

Figure 8-9. DSN Advanced Options dialog box

Once the DSN is successfully created, it should appear in the *System DSN* list, as shown in Figure 8-10.

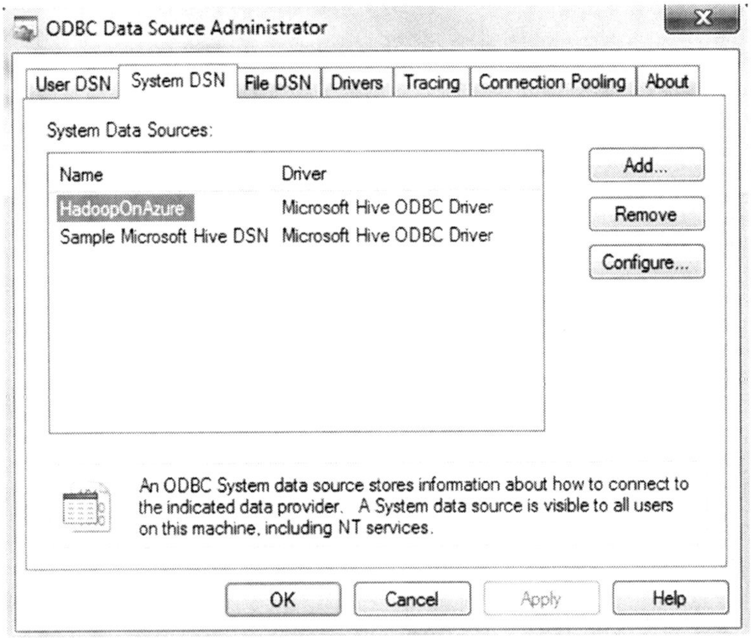

Figure 8-10. *The HadoopOnAzure System DSN*

■ **Note** When you install the ODBC driver, a sample DSN is automatically created, called *Sample Microsoft Hive DSN*.

Connecting to the HDInsight Emulator

There are a few differences between connecting to Windows Azure HDInsight service and the single-node HDInsight Emulator on premises using the ODBC driver. You are not required to provide any user name or password to connect to the emulator. The other two key differences between connecting to the emulator and connecting to the Azure service are:

- **Port Number**   The ODBC driver connects to the emulator using port *10001*.

- **Authentication Mechanism**   The mechanism used is *Windows Azure HDInsight Emulator*.

Figure 8-11 shows the configuration screen of the ODBC DSN when connecting to the HDInsight Emulator.

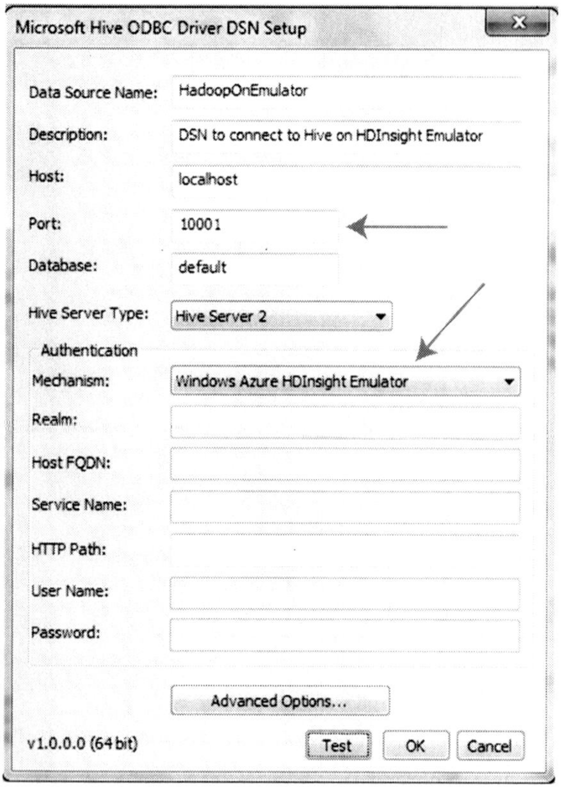

Figure 8-11. *Connecting to Windows Azure HDInsight Emulator*

Configuring a DSN-less Connection

Using a DSN requires you to preregister the data source using the Windows ODBC Data Source Administrator. You'll then be able to reference this DSN entry by name from any client application connection. A DSN essentially creates an alias for your data source—you can change where the DSN is pointing and your applications will continue to work. However, the downside to this approach is that you'll need to make sure the DSN exists on all machines that will be running your applications. The alternate way of establishing a connection without a DSN is to use a connection string in your application. The advantage of using a connection string is that you don't have to pre-create the DSN on the systems that will be running your application.

The connection string parameters can be a little tricky, but this is a preferred approach because it removes the external (DSN) dependency. Also note that the same connection string works for both 32-bit and 64-bit execution modes. So you can avoid creating multiple DSNs—you just need to ensure that both versions of the ODBC driver are installed. Table 8-1 summarizes the connection string attribute you need to set to create a DSN-less connection to Hive using the Microsoft Hive ODBC driver.

Table 8-1. *Connection string to hive*

Field	Description
Driver	Name of the driver: {Microsoft Hive ODBC Driver}.
Host	DNS hostname of your cluster.
Port	Connection port: The Azure HDInsight Service is 443, and the Azure HDInsight Emulator is 10001.
Schema	Default database schema.
RowsFetchedPerBlock	Number of rows fetched per block. The recommendation is to keep it at 10,000.
HiveServerType	The HDInsight default is 2.
AuthMech	Authentication mechanism: You'll want to use a value of 6, which maps to using the username and password you specified when the cluster was created, or a value of 3 to connect to the Emulator.
DefaultStringColumnLength	The default length for STRING columns.

A sample connection string using an ODBC DSN named HDISample should look like this:

```
Provider=MSDASQL.1;Password=**********;Persist Security Info=True;User ID=admin;

Data Source=HDISample;Initial Catalog=HIVE
```

Note that there are only a few mandatory parameters that need to be passed in the connection string, such as Provider, Data Source, User ID, and Password. The rest of the details, like Port Number and Authentication Mechanism, are embedded in the DSN itself and should be correctly provided while creating the DSN.

Summary

Hive acts as a data warehouse on top of HDFS (WASB, in case of HDInsight), providing an easy and familiar SQL-like query language called HQL to fetch the underlying data. HQL queries are broken down into MapReduce code internally, relieving the end user from writing complex MapReduce code. The Hive ODBC driver acts as an interface between client consumers and HDInsight, enabling access from any tool supporting ODBC. In this chapter, you learned about creating and working with Hive tables, as well as configuring and connecting to Azure HDInsight Service and Azure HDInsight Emulator using the Microsoft Hive ODBC driver. You also learned to create a DSN-less connection to HDInsight for client applications to connect using a connection string.

CHAPTER 9

Consuming HDInsight from Self-Service BI Tools

Self-service (Business Intelligence) BI is the talk of the town at the moment. As the term suggests, self-service BI is a concept through which you can perform basic data analysis and extract intelligence out of that data with easy-to-use tools, without needing to hire a suite of BI experts or implement a data warehouse solution. Self-service BI is certainly a trend toward the consumerization of IT and BI. The trend is that an individual or even a really small-scale and growing company can afford BI to implement a better decision-making process. This chapter will focus on the various self-service BI tools available from Microsoft that provide strong integration with HDInsight and help in the following analytics and reporting processes:

- PowerPivot
- Power View
- Power BI

PowerPivot Enhancements

With SQL Server 2012, Microsoft has enhanced the data-analysis capabilities of PowerPivot for both the client-side component (PowerPivot for Excel) and the server-side component (PowerPivot for SharePoint) to provide enhanced self-service BI functionality to all Microsoft Office users. The new enhancements in PowerPivot help users integrate data from multiple sources more easily, create reports and analytical applications faster, and share and collaborate on insights more easily using the familiar environments of Excel and SharePoint.

PowerPivot comes as an add-in to Excel 2013 and Excel 2010 that allows business users to work with data from any source and syndication, including Open Data Protocol (ODATA) feeds, to create business models and integrate large amounts of data directly into Excel workbooks. Sophisticated workbooks can be built using Excel only, or using the PowerPivot model as a source of data from other BI tools. These BI tools can include third-party tools as well as the new Power View capability (discussed later in this chapter) to generate intelligent and interactive reports. These reports can be published to SharePoint Server and then shared easily across an organization.

The following section explains how to generate a PowerPivot data model based on the *stock_analysis* Hive table created earlier using the Microsoft Hive ODBC Driver. We have used Excel 2013 for the demos. Open a new Excel worksheet, and make sure you turn on the required add-ins for Excel as shown in Figure 9-1. You'll need those add-ins enabled to build the samples used throughout this chapter. Go to *File ➤ Options ➤ Add-ins.* In the Manage drop-down list, click *COM Add-ins ➤ Go* and enable the add-ins.

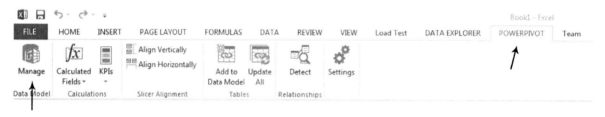

Figure 9-1. Enabling the Excel add-ins

■ **Note**   PowerPivot is also supported in Excel 2010. Power View and Power Query are available only in Excel 2013.

To create a PowerPivot model, open Excel, navigate to the *POWERPIVOT* ribbon, and click on *Manage* as shown in Figure 9-2.

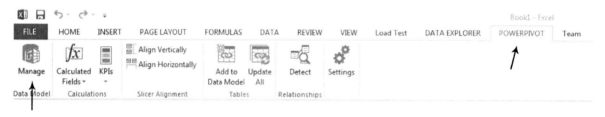

Figure 9-2. PowerPivot for Excel 2013

Clicking on the *Manage* icon will bring up the PowerPivot for Excel window where you need to configure the connection to Hive. Click on *Get External Data*, and select *From other Sources* as shown in Figure 9-3.

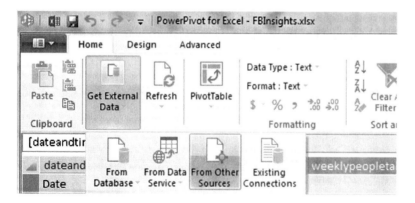

Figure 9-3. PowerPivot for Excel

Because you are using the Hive ODBC driver, choose *Others (OLEDB/ODBC)* and click *Next* on the *Table Import Wizard* as shown in Figure 9-4.

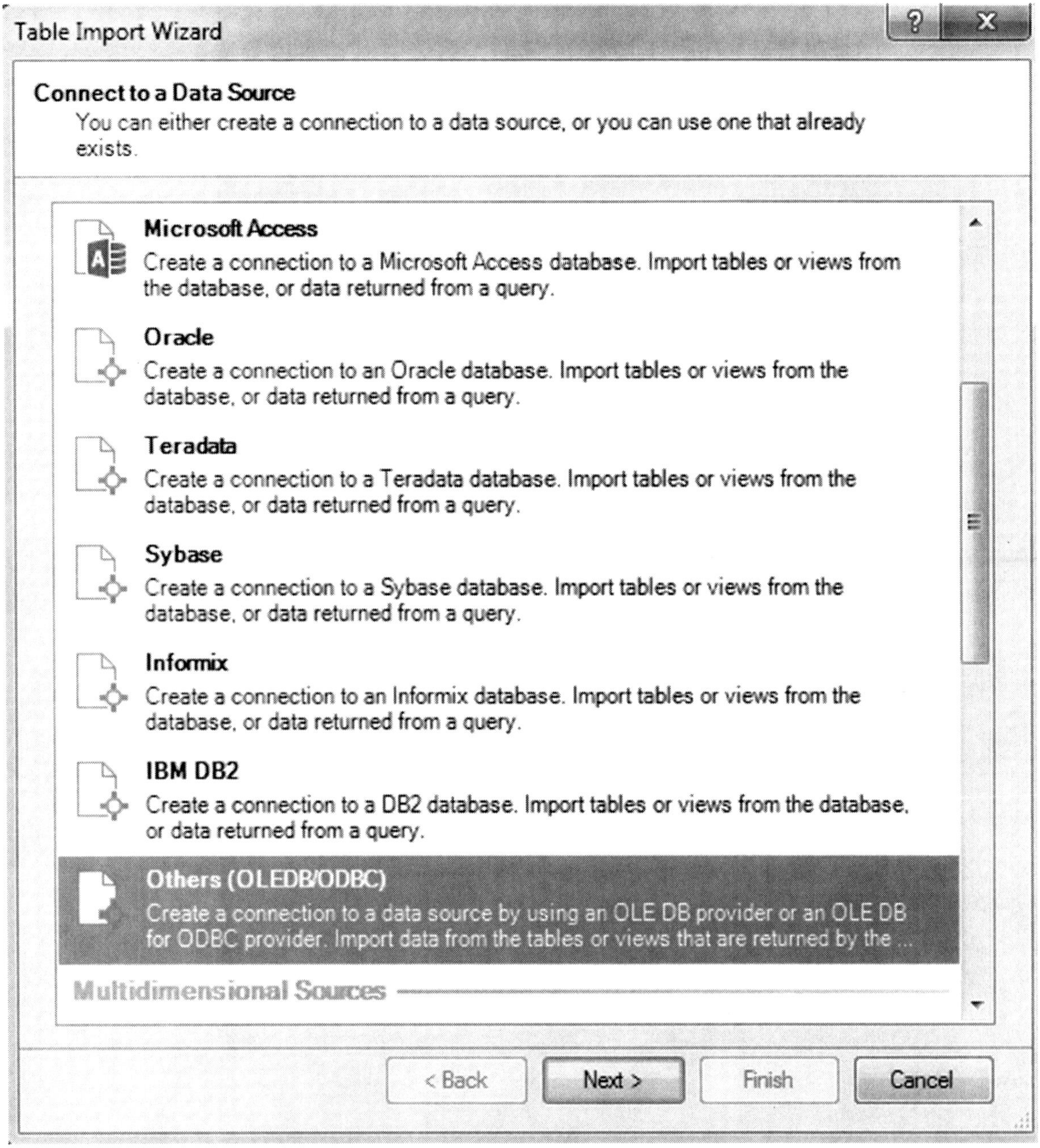

Figure 9-4. *Selecting a provider*

The next screen in the wizard accepts the connection string for the data source. You can choose to build the connection string instead of writing it manually. So click on the *Build* button to bring up the Data Link file, where you can select the *HadoopOnAzure* DSN we created earlier and provide the correct credentials to access the HDInsight cluster. Make sure to select *Allow saving password* so that the password is retained in the underlying PowerPivot *Table Import Wizard*. Also, verify that *Test Connection* succeeds as shown in Figure 9-5. If you provide all the details correctly, you should also be able to enumerate the default database HIVE in the *Enter the initial catalog to use* drop-down list.

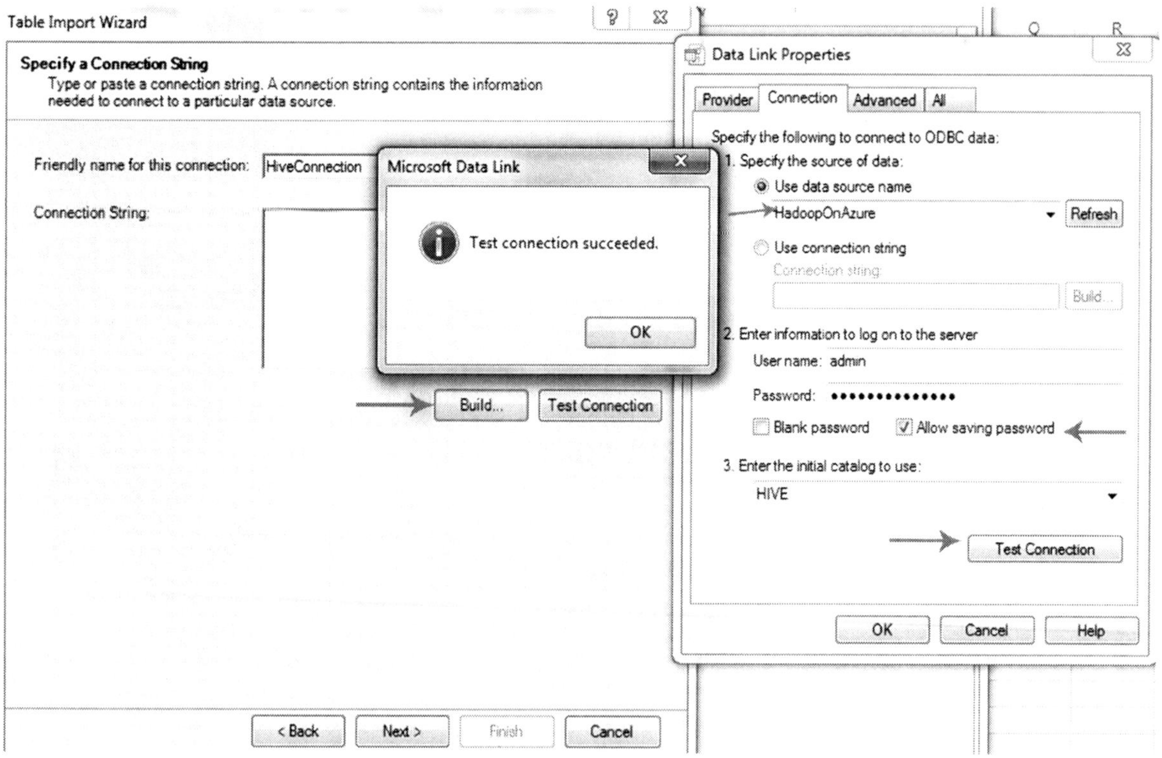

Figure 9-5. *Configuring the connection string*

The *Table Import Wizard* dialog should be populated with the appropriate connection string as shown in Figure 9-6. Click on *Next*.

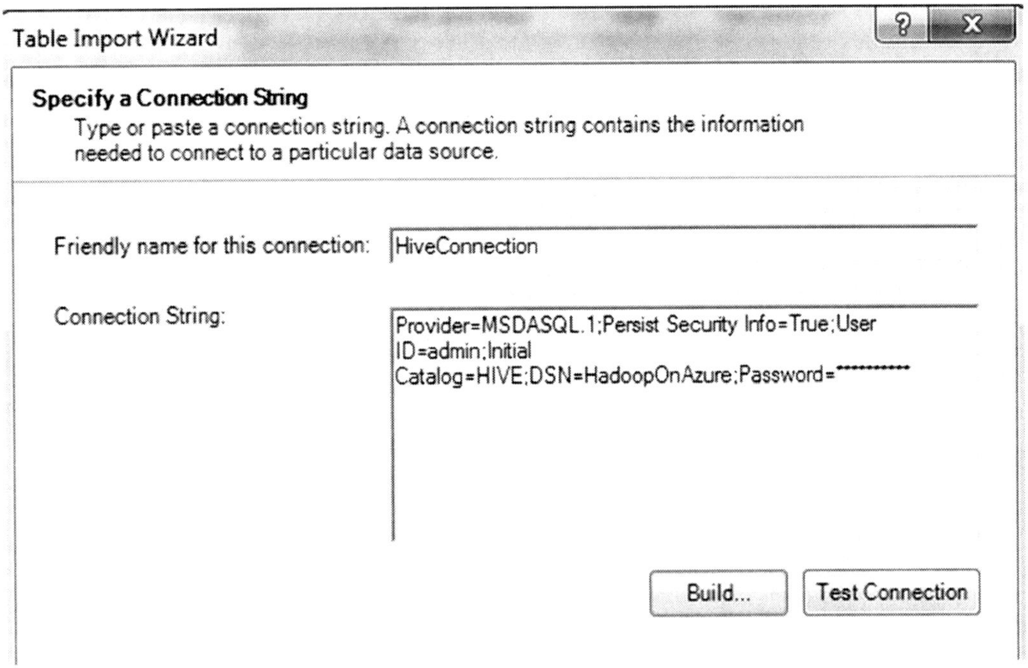

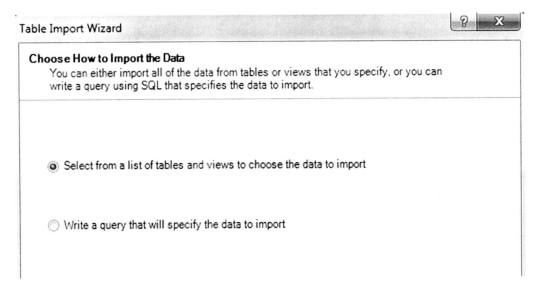

Figure 9-6. *Configuring the connection string*

We are going to choose to import from the Hive table directly, but we can also write a query (HiveQL) to fetch the data, as shown in the Figure 9-7.

Table Import Wizard

Choose How to Import the Data
You can either import all of the data from tables or views that you specify, or you can write a query using SQL that specifies the data to import.

◉ Select from a list of tables and views to choose the data to import

○ Write a query that will specify the data to import

Figure 9-7. *Select the table or write the query*

Select the *stock_analysis* table, and click *Finish* to complete the configuration as shown in Figure 9-8.

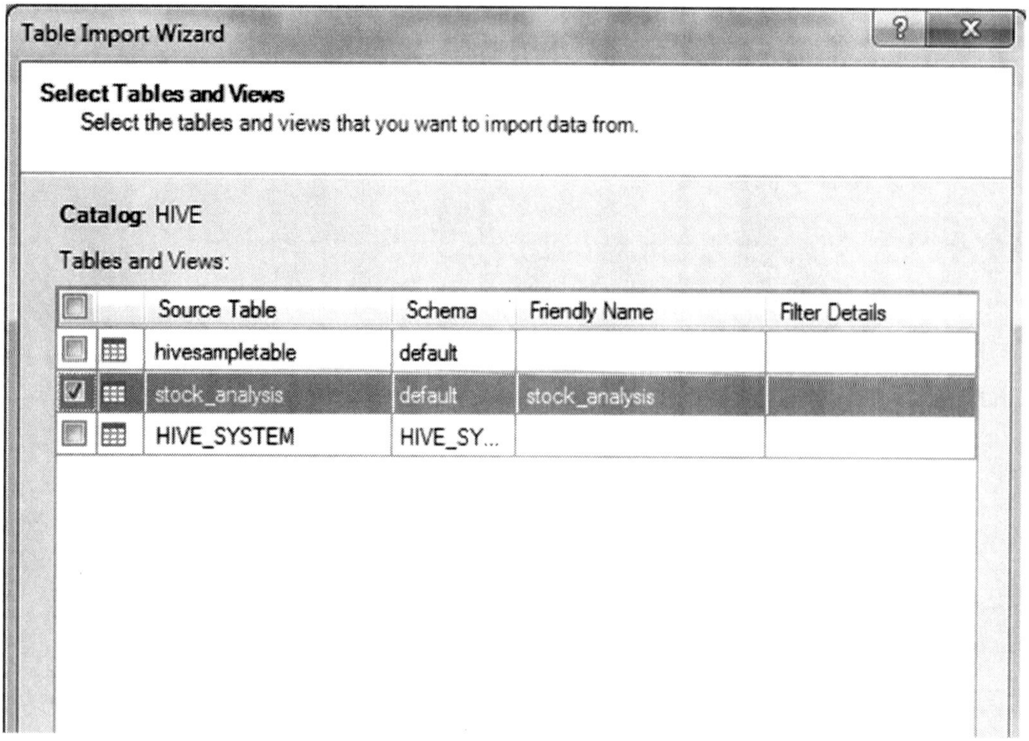

Figure 9-8. *Selecting the table*

The Hive table with all the rows should get successfully loaded in the PowerPivot model as shown in Figure 9-9.

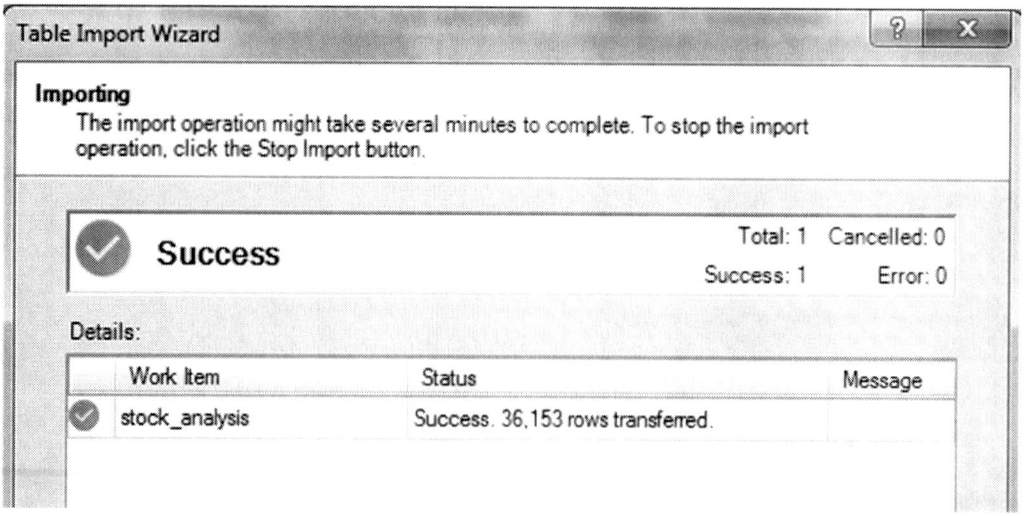

Figure 9-9. *Finishing the import*

Close the *Table Import Wizard*. You should see the PowerPivot model populated with data from the `stock_analysis` table in Hive as shown in Figure 9-10.

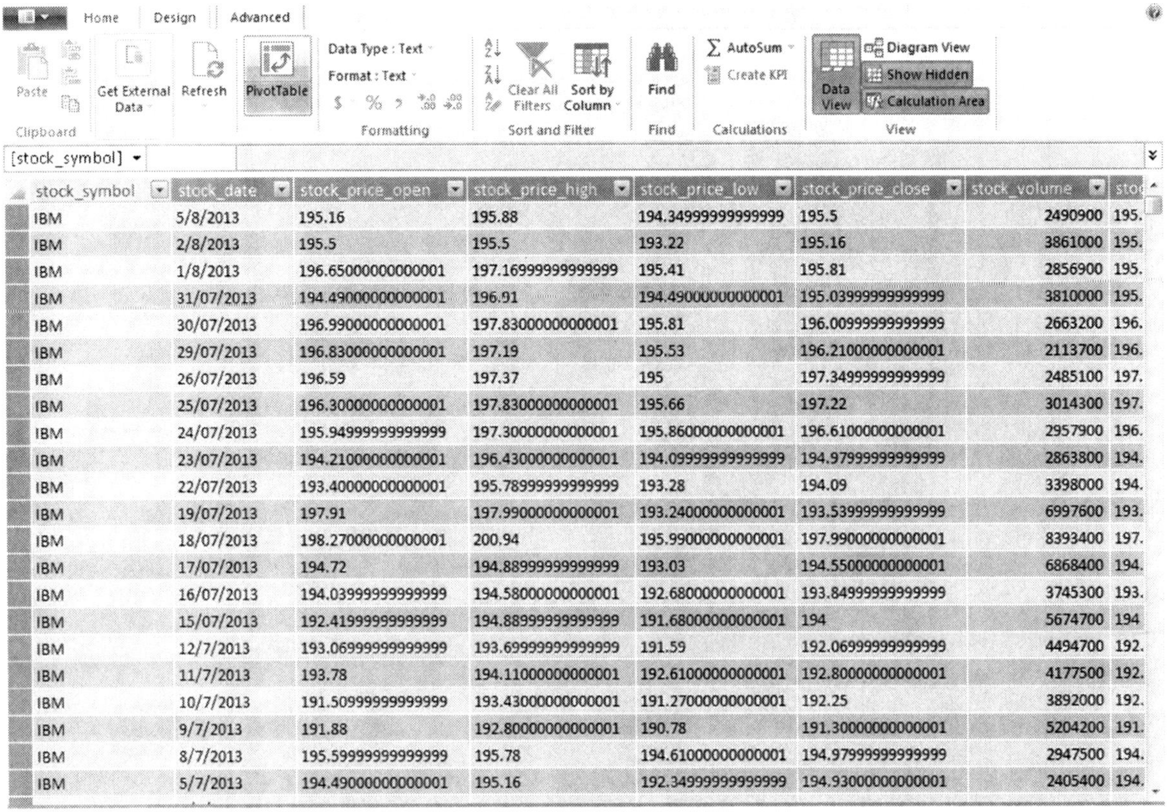

stock_symbol	stock_date	stock_price_open	stock_price_high	stock_price_low	stock_price_close	stock_volume	sto
IBM	5/8/2013	195.16	195.88	194.349999999999	195.5	2490900	195.
IBM	2/8/2013	195.5	195.5	193.22	195.16	3861000	195.
IBM	1/8/2013	196.650000000001	197.169999999999	195.41	195.81	2856900	195.
IBM	31/07/2013	194.490000000001	196.91	194.490000000001	195.039999999999	3810000	195.
IBM	30/07/2013	196.990000000001	197.830000000001	195.81	196.009999999999	2663200	196.
IBM	29/07/2013	196.830000000001	197.19	195.53	196.210000000001	2113700	196.
IBM	26/07/2013	196.59	197.37	195	197.349999999999	2485100	197.
IBM	25/07/2013	196.300000000001	197.830000000001	195.66	197.22	3014300	197.
IBM	24/07/2013	195.949999999999	197.300000000001	195.860000000001	196.610000000001	2957900	196.
IBM	23/07/2013	194.210000000001	196.430000000001	194.099999999999	194.979999999999	2863800	194.
IBM	22/07/2013	193.400000000001	195.789999999999	193.28	194.09	3398000	194.
IBM	19/07/2013	197.91	197.990000000001	193.240000000001	193.539999999999	6997600	193.
IBM	18/07/2013	198.270000000001	200.94	195.990000000001	197.990000000001	8393400	197.
IBM	17/07/2013	194.72	194.889999999999	193.03	194.550000000001	6868400	194.
IBM	16/07/2013	194.039999999999	194.580000000001	192.680000000001	193.849999999999	3745300	193.
IBM	15/07/2013	192.419999999999	194.889999999999	191.680000000001	194	5674700	194
IBM	12/7/2013	193.069999999999	193.699999999999	191.59	192.069999999999	4494700	192.
IBM	11/7/2013	193.78	194.110000000001	192.610000000001	192.800000000001	4177500	192.
IBM	10/7/2013	191.509999999999	193.430000000001	191.270000000001	192.25	3892000	192.
IBM	9/7/2013	191.88	192.800000000001	190.78	191.300000000001	5204200	191.
IBM	8/7/2013	195.599999999999	195.78	194.610000000001	194.979999999999	2947500	194.
IBM	5/7/2013	194.490000000001	195.16	192.349999999999	194.930000000001	2405400	194.

Figure 9-10. *The PowerPivot data model*

Change the data type of the column *stock_date* to `Date` as shown in Figure 9-11.

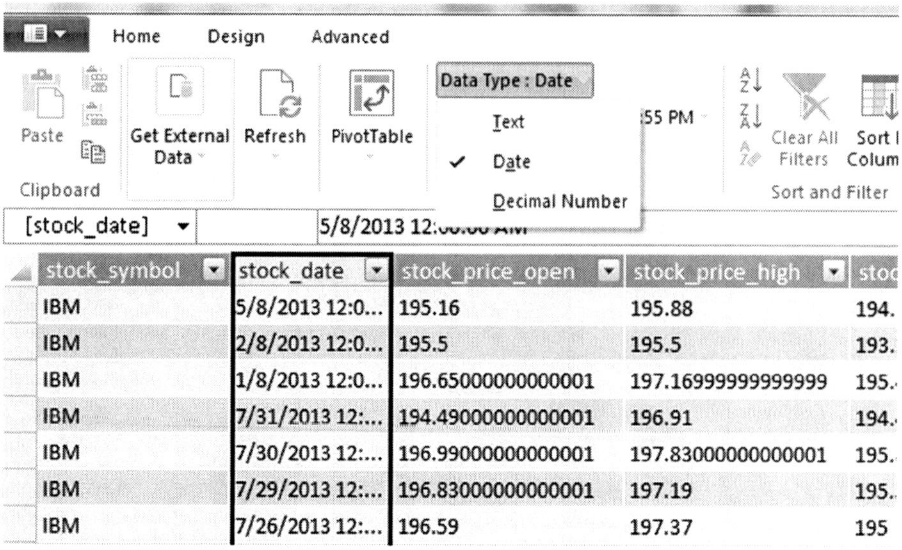

Figure 9-11. *Changing the stock_date to the* Date *data type*

Select the columns from *stock_price_open* to *stock_price_close*, and set their data type to Decimal as shown in Figure 9-12.

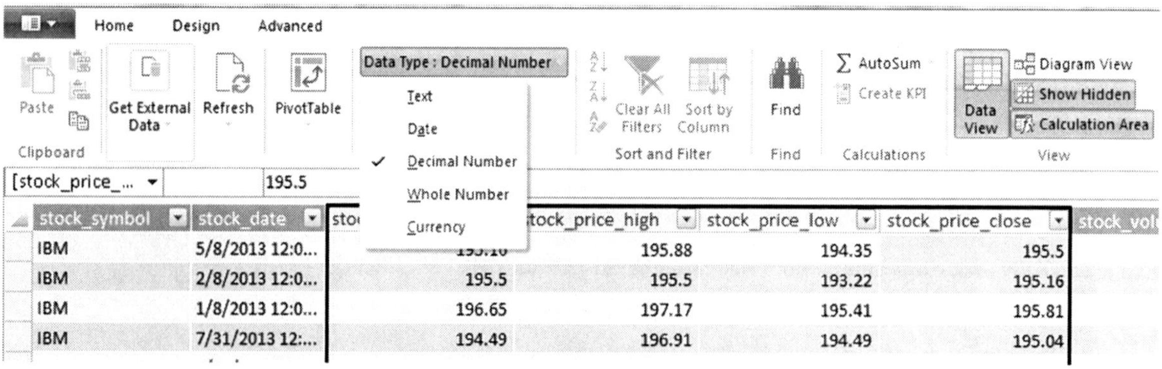

Figure 9-12. *Changing columns to the* Decimal *data type*

Select the *stock_price_adj_close column*, and set its type to Decimal as well. Next, import the *DimDate* table from *AdventureWorksDWH* database in SQL Server to be able to create a date hierarchy. Click on *Get External Data* ➤ *From Other Sources* ➤ *Microsoft SQL Server*, and provide the SQL Server connection details as shown in Figure 9-13.

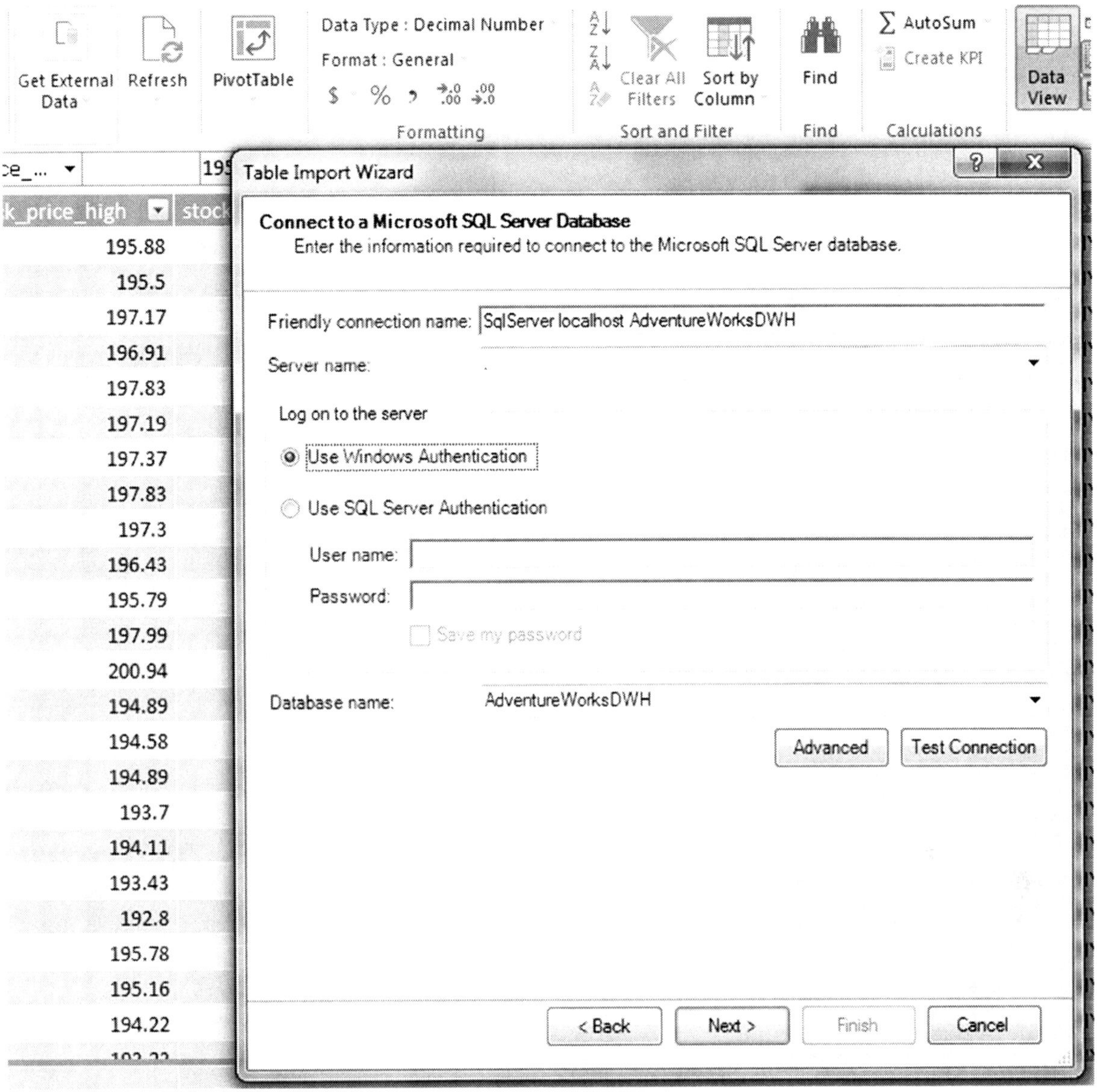

Figure 9-13. *Getting data from the AdventureWorksDWH database in SQL Server*

Click on *Next*, choose to import from table directly, and click on *Next* again. Select the *DimDate* table from the available list of tables to import in the model as shown in Figure 9-14.

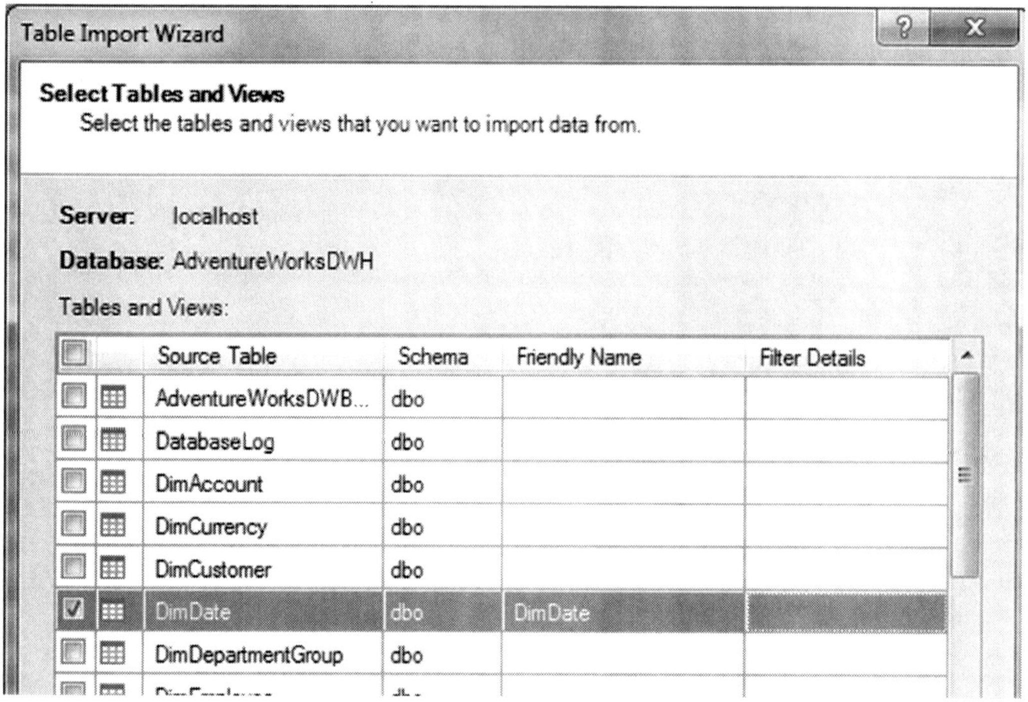

Figure 9-14. Choosing the DimDate table

If you do not have the AdventureWorksDWH database, you can download it from the following link:

http://msftdbprodsamples.codeplex.com/releases/view/55330

■ **Note** You will see a lot of sample database files available for download when you link to the site just mentioned. For this chapter, download the file that says AdventureWorksDW2012 Data File. After the download is complete, make sure you attach the database to your SQL Server instance. You can do so using the SQL Server *Attach Databases* wizard or simply executing the following SQL statement:

```
EXEC sp_attach_single_file_db @dbname = 'AdventureWorksDWH', @filename =
'<path>\AdventureWorksDW2012_Data.mdf'
```

Once the import of the DimDate table is done, your PowerPivot data model will have two tables loaded in it. The tables are named stock_analysis and DimDate.

Creating a Stock Report

Once the two tables are loaded into the PowerPivot model, click on *Diagram View* and connect the *DimDate* table with the *stock_analysis* table using *Full Date Alternate Key* and *stock_date* as shown in Figure 9-15. (Drag *stock_date* to the *Full Date Alternate Key* column.)

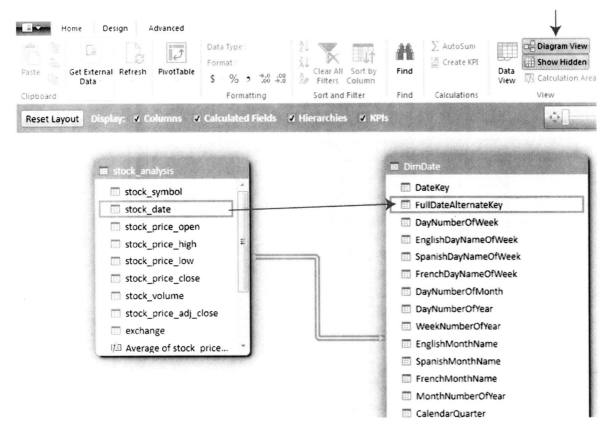

Figure 9-15. *Creating the relation*

Click on the *Create Hierarchy* button in the DimDate table. Create a hierarchy for CalendarYear, CalendarQuarter, EnglishMonthName, and FullDateAlternateKey as shown in Figure 9-16. (Drag and drop these four columns under the hierarchy's HDate value.)

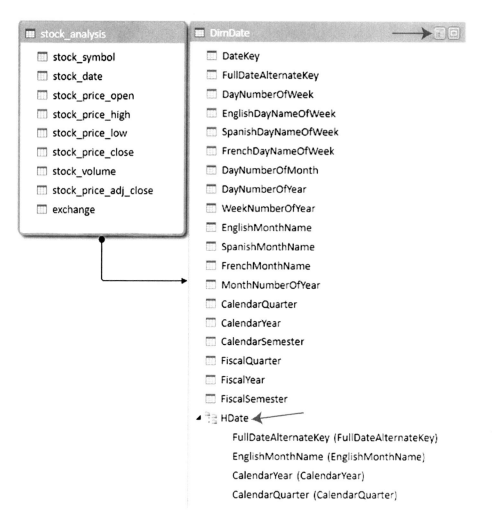

Figure 9-16. *Creating the hierarchy*

Next, go back to the *Data View* and create measures for your stock table. Select *stock_price_open, stock_price_high, stock_price_low, stock_price_close,* and choose *Average* under *AutoSum.* Doing that will create measures with average calculations as shown in Figure 9-17.

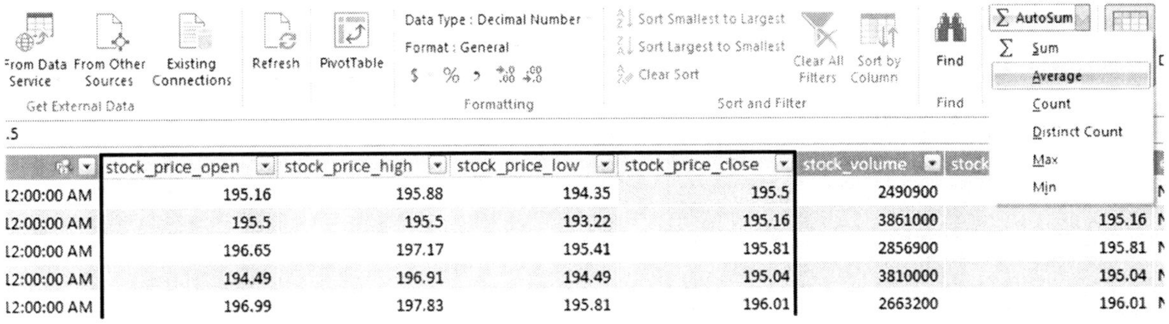

Figure 9-17. *Creating measures for stocks*

Go ahead and add another measure for *stock_volume,* but this time make *Sum* the aggregation function. Once the measures are created, click on *PivotTable➤PivotChart* as shown in Figure 9-18. That will open a new worksheet with a chart.

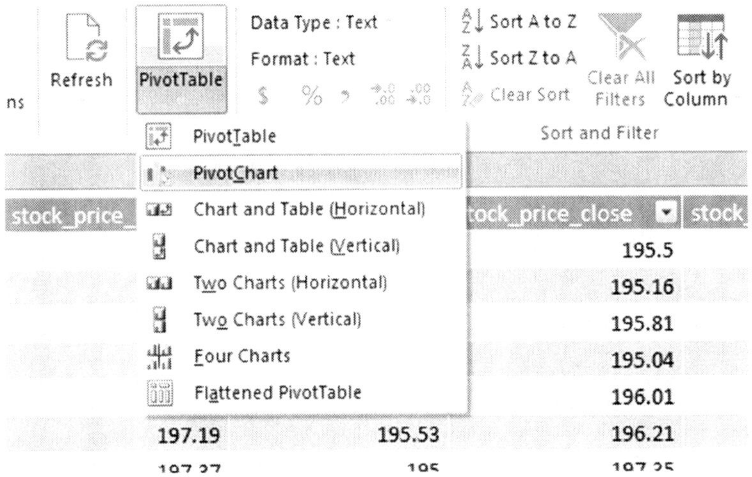

Figure 9-18. *Creating a PivotChart*

Once the new worksheet with the data models is open, drag and drop *stock_symbol* to *Legend (Series).* Then drag *HDate* to *Axis (Category)* and *Average of Stock Price Close* to *Values* as shown Figure 9-19.

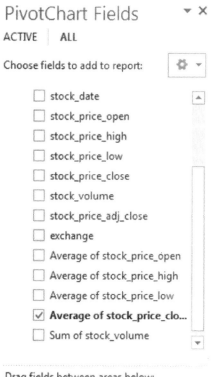

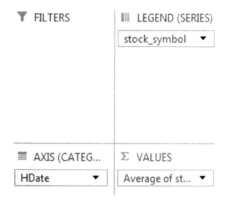

Figure 9-19. *Designing the PivotChart*

You should be able to see a graphical summary of the closing price of the stocks of the companies over a period of time as shown in Figure 9-20.

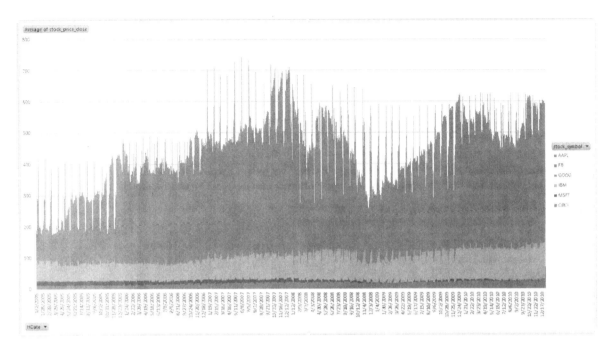

Figure 9-20. The stock summary chart

In the next section, you will see how you can use Power View to consume the PowerPivot data model and quickly create intelligent and interactive visualizations out of the stock market data.

Power View for Excel

Power View is a feature of Microsoft Excel 2013, and it's also a feature of Microsoft SharePoint 2013 as part of the SQL Server 2012 Service Pack 1 Reporting Services Add-in for Microsoft SharePoint Server 2013 Enterprise Edition. Power View in Excel 2013 and Power View in SharePoint 2013 both provide an interactive data exploration, visualization, and presentation experience for all skill levels, and they have similar features for designing Power View reports.

This chapter shows a sample Power View report based on the *stock_analysis* table's data in Hive to give you a quick look at the powerful visualization features from the surface level. Details about how to design a Power View report as well as details about Power View integration with SharePoint are outside the scope of this book. Neither topic is discussed in depth.

■ **Note**   Power View is supported only in Excel 2013.

To create a Power View report based on the PowerPivot data model created earlier, open the workbook with the PowerPivot model, click on the *Insert* ribbon in Excel, and select *Power View* as shown in Figure 9-21.

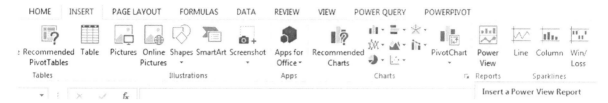

Figure 9-21. *Launching Power View for Excel*

This launches a new Power View window with the PowerPivot model already available to it. With the Power View window open, you are now ready to create a report, chart, or other visualization. Here are the steps to follow:

1. Once Power View opens, click on the chart to select and highlight it. Drag and drop *Average of stock_price_close* into the *fields* section.

2. Click the *Line Chart* graph in *Design* ribbon to switch to the chart and expand the graph to fit it to the canvas.

3. Change the title to *Stock Comparison*.

4. Drag *Hdate* to the *Filters* field in the report.

5. Drag *exchange* to the *Tile By* column.

6. Drag *FullDateAlternateKey* to *Axis*.

7. Drag *stock_symbol* to *Legend*.

Once the design is complete, you should be able to see the Power View report comparing the different stock prices in a line chart. It is categorized based on the NASDAQ and NYSE, and it gives you a visualization of the stock prices with just a few clicks. Your Power View report should now look like Figure 9-22.

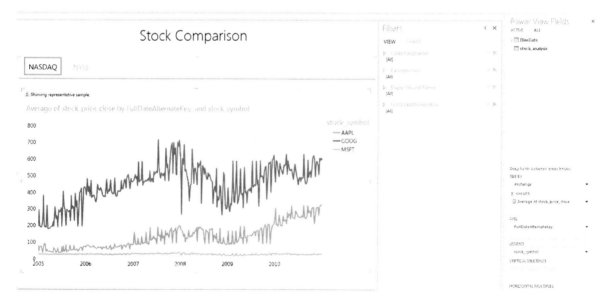

Figure 9-22. *The Stock Comparison report*

Power BI: The Future

Power BI for Office provides users with powerful new ways to work with data in Excel from a variety of data sources. It lets you easily search, discover, and access data within and outside your organization and, with just a few clicks, shape, transform, analyze, and create stunning interactive visualizations out of the data. These visualizations uncover hidden insights you can share, and you can collaborate from anywhere, on any device. In this section, you will look at two offerings in the Power BI suite:

- Power Query

- Power Map

Power Query is a *mash-up* tool designed to integrate data from heterogeneous data sources through an easy-to-use, graphical interface. It is available as an add-in to Excel after you download it from the following web site:

```
http://www.microsoft.com/en-us/download/details.aspx?id=39379
```

Power Map is an upcoming offering that previously was known as GeoFlow. Power Map can be used together with Power Query to create stunning, three-dimensional visualizations of coordinates plotted over Bing maps. Learn more about Power Map from the following article on Microsoft Developer Network:

```
http://blogs.msdn.com/b/powerbi/archive/2013/07/07/getting-started-with-pq-and-pm.aspx
```

In this section, you will use Power Query to connect to your Windows Azure HDInsight Service and load data from HDFS to your Excel worksheet. Begin from the Excel toolbar ribbon by clicking *Power Query* ➤ *From Other Sources* ➤ *From Windows Azure HDInsight* as shown in Figure 9-23.

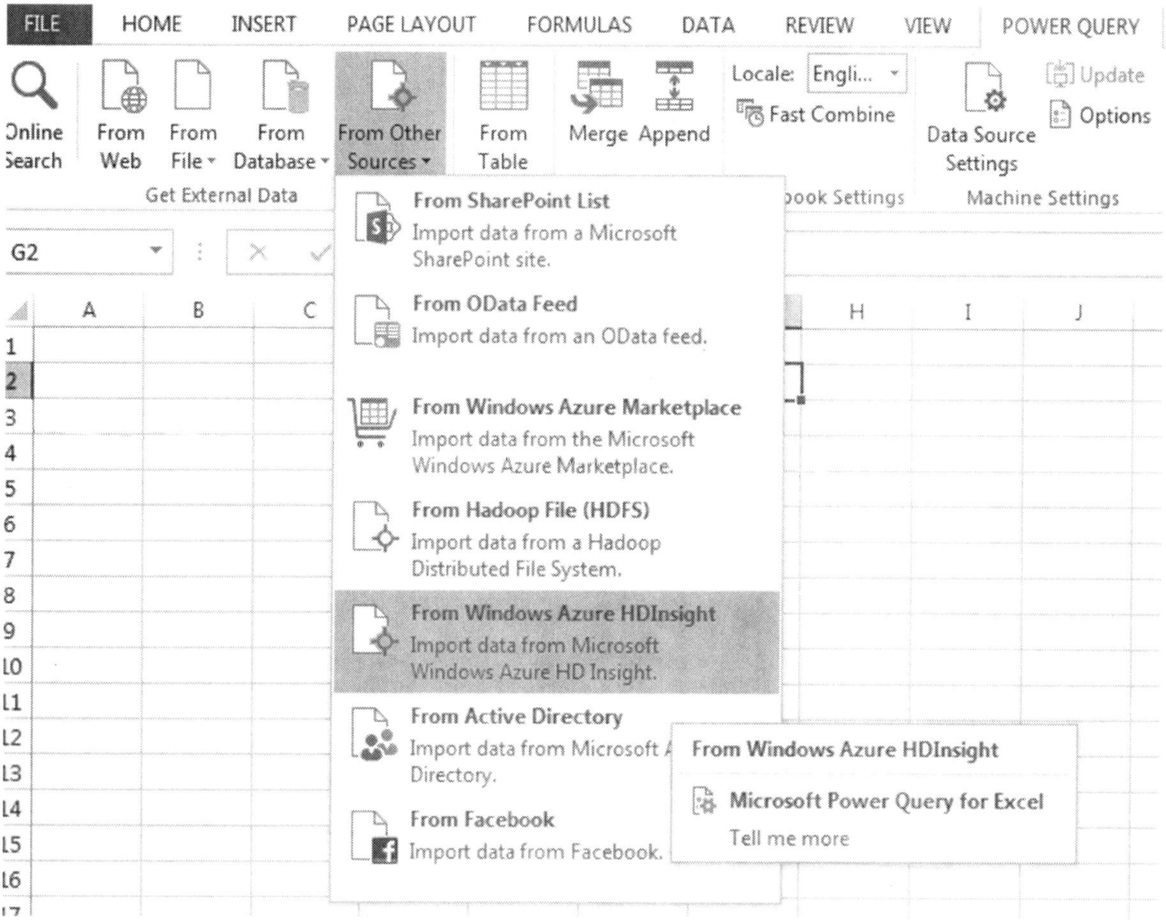

Figure 9-23. *Connecting Power Query to Azure HDInsight*

Excel will prompt you for the cluster storage account name as shown in Figure 9-24. Provide your storage account name and click *OK*.

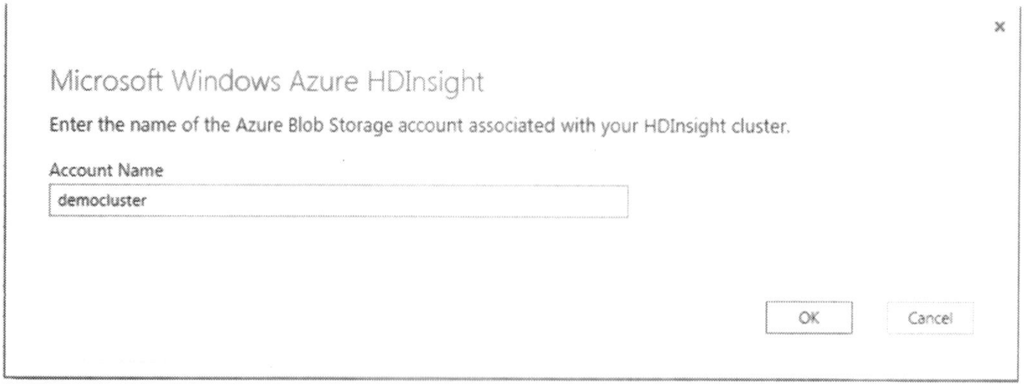

Figure 9-24. *Provide a cluster storage account name*

When you connect for the first time, Excel will also prompt you for your storage account key. Enter that key and click on *Save*. Click on *Edit Query* to load the *Query Editor* screen, where you can specify your filter criteria. Expand the drop-down list under the *Name* column, and filter only the rows that have .csv files, as shown in Figure 9-25.

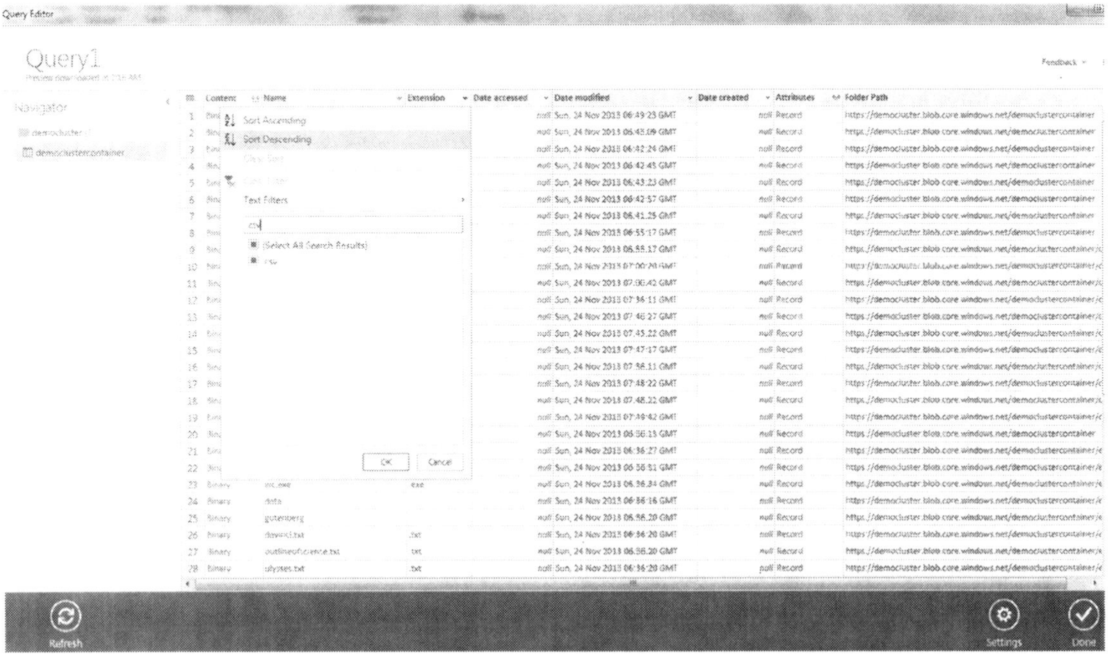

Figure 9-25. *Filtering .csv files*

Click on *OK,* and then click the *Binary* link under the *Content* column in Figure 9-25. Doing so will load the data from the *.csv* file into the *Query Editor*. You can rename the columns to more meaningful names before importing them into the Excel worksheet, as illustrated in Figure 9-26.

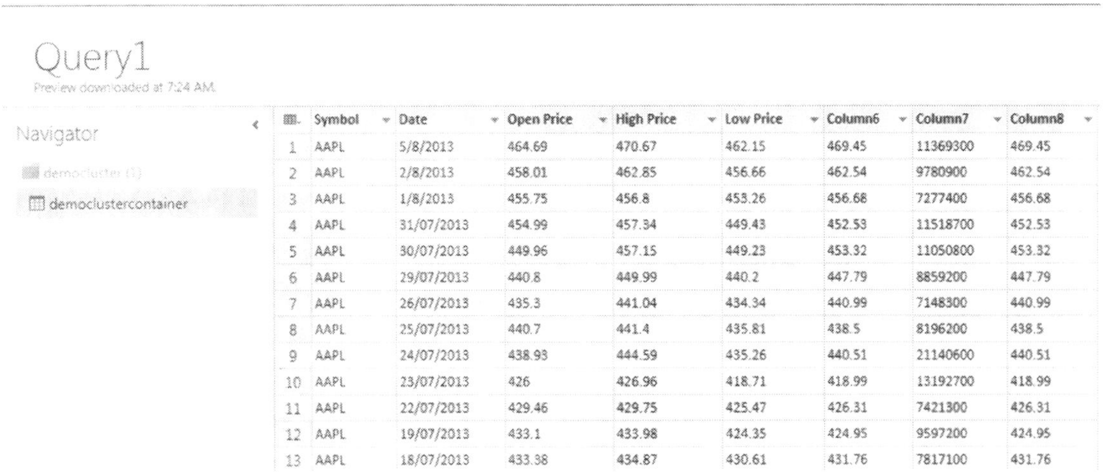

Figure 9-26. *Formatting the data*

Click on *Apply and Close,* and the data will be imported to Excel. You can see the total number of rows downloaded from your blob storage and even load the data into a data model as shown in Figure 9-27.

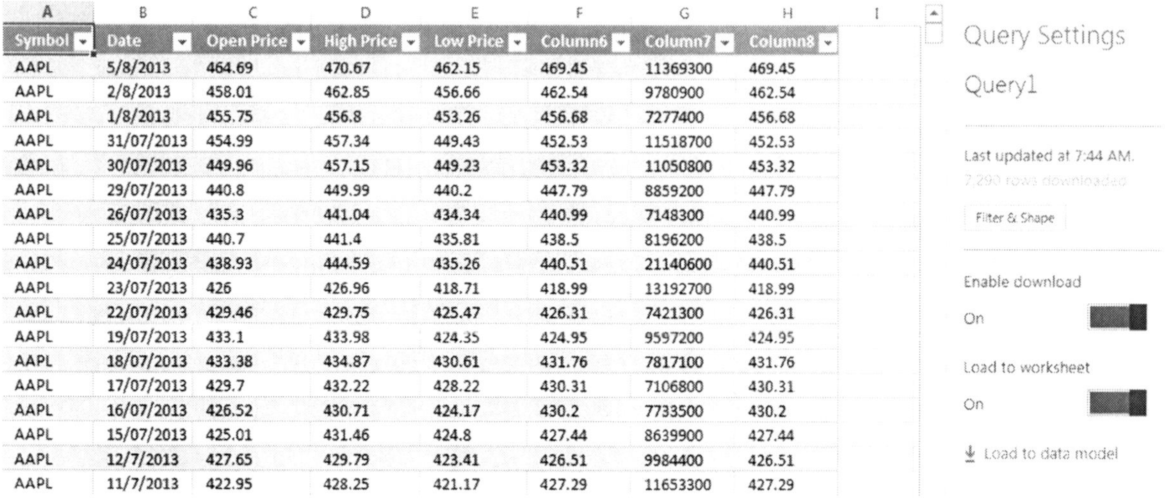

Symbol	Date	Open Price	High Price	Low Price	Column6	Column7	Column8
AAPL	5/8/2013	464.69	470.67	462.15	469.45	11369300	469.45
AAPL	2/8/2013	458.01	462.85	456.66	462.54	9780900	462.54
AAPL	1/8/2013	455.75	456.8	453.26	456.68	7277400	456.68
AAPL	31/07/2013	454.99	457.34	449.43	452.53	11518700	452.53
AAPL	30/07/2013	449.96	457.15	449.23	453.32	11050800	453.32
AAPL	29/07/2013	440.8	449.99	440.2	447.79	8859200	447.79
AAPL	26/07/2013	435.3	441.04	434.34	440.99	7148300	440.99
AAPL	25/07/2013	440.7	441.4	435.81	438.5	8196200	438.5
AAPL	24/07/2013	438.93	444.59	435.26	440.51	21140600	440.51
AAPL	23/07/2013	426	426.96	418.71	418.99	13192700	418.99
AAPL	22/07/2013	429.46	429.75	425.47	426.31	7421300	426.31
AAPL	19/07/2013	433.1	433.98	424.35	424.95	9597200	424.95
AAPL	18/07/2013	433.38	434.87	430.61	431.76	7817100	431.76
AAPL	17/07/2013	429.7	432.22	428.22	430.31	7106800	430.31
AAPL	16/07/2013	426.52	430.71	424.17	430.2	7733500	430.2
AAPL	15/07/2013	425.01	431.46	424.8	427.44	8639900	427.44
AAPL	12/7/2013	427.65	429.79	423.41	426.51	9984400	426.51
AAPL	11/7/2013	422.95	428.25	421.17	427.29	11653300	427.29

Query Settings

Query1

Last updated at 7:44 AM.
7,290 rows downloaded

Filter & Shape

Enable download
On

Load to worksheet
On

↓ Load to data model

Figure 9-27. Importing data using Power Query

■ **Note**   Power Query can directly fetch HDFS data and place it in your Excel worksheet in a formatted way. There is no need to access the data through Hive, and there is no dependency on the Microsoft Hive ODBC Driver as with PowerPivot or Power View.

You can repeat the preceding steps to connect to different types of data sources and integrate them in your Excel sheet. The integrated data can then be filtered and shaped to create curated data models targeting specific business requirements.

Summary

In this chapter, you learned how to integrate Microsoft self-service BI tools with HDInsight to consume data and generate powerful visualizations of the data. With the paradigm shifts in technology, the industry is trending toward an era in which Information Technology will be a consumer product. An individual person will be able to visualize the insights he needs to an extent from a client-side add-in like Power View. You also had a peep into the Power BI tools that are available from Microsoft to provide data mash-ups and 3-D visualizations of your data. These self-service BI tools provide the capability of connecting and talking to a wide variety of data sources seamlessly and creating in-memory data models that combine the data from these diverse sources for powerful reporting.

■ ■ ■

## CHAPTER 10

Integrating HDInsight with SQL Server Integration Services

Microsoft SQL Server is a complete suite of tools that include a relational database management system (RDBMS), multidimensional online analytical processing (OLAP) and tabular database engines, a brokering service, a scheduling service (SQL Agent), and many other features. As discussed in Chapter 1, it has become extremely important these days to integrate data between different sources. The advantage that SQL Server brings is that it offers a powerful Business Intelligence (BI) stack, which provides rich features for data mining and interactive reporting. One of these BI components is an Extract, Transform, and Load (ETL) tool called *SQL Server Integration Services (SSIS)*. ETL is a process to extract data, mostly from different types of systems, transform it into a structure that's more appropriate for reporting and analysis and finally load it into the database. SSIS, as an ETL tool offers the ability to merge structured and unstructured data by importing Hive data into SQL Server and apply powerful analytics on the integrated data. Throughout the rest of this chapter, you will get a basic lesson on how SSIS works and create a simple SSIS package to import data from Hive to SQL Server.

SSIS as an ETL Tool

The primary objective of an ETL tool is to be able to import and export data to and from heterogeneous data sources. This includes the ability to connect to external systems, as well as to transform or clean the data while moving the data between the external systems and the databases. SSIS can be used to import data to and from SQL Server. It can even be used to move data between external non-SQL systems without requiring SQL Server to be the source or the destination. For instance, SSIS can be used to move data from an FTP server to a local flat file.

SSIS also provides a workflow engine to automate various tasks (data flows, task executions, and so forth) that are executed in an ETL job. An SSIS package execution can itself be one step that is part of an SQL Agent job, and SQL Agent can run multiple jobs independent of each other.

An SSIS solution consists of one package or more, each containing a control flow to perform a sequence of tasks. Tasks in a control flow can include calls to web services, FTP operations, file-system tasks, the automation of command-line commands, and others. In particular, a control flow usually includes one or more data-flow tasks, which encapsulate an in-memory, buffer-based pipeline of data from a source to a destination, with transformations applied to the data as it flows through the pipeline. An SSIS package has one control flow and as many data flows as necessary. Data-flow execution is dictated by the content of the control flow.

Detailed discussion on SSIS and its components are outside the scope of this book. In this chapter, I assume you are familiar with basic SSIS package development using Business Intelligence Development Studio (BIDS, in SQL Server 2005/2008/2008 R2) or SQL Server Data Tools (in SQL Server 2012). If you are a beginner in SSIS, I recommend that you read one of the many good, introductory SSIS books available as a pre-requisite. In the rest of this chapter, we will focus on how to consume Hive data from SSIS using the Hive Open Database Connectivity (ODBC) driver.

The pre-requisites to developing the package shown in this chapter are SQL Server Data Tools (which comes as a part of SQL Server 2012 Client Tools and Components) and the 32-Bit Hive ODBC Driver installed. You will also need either an on-premise HDInsight Emulator or a subscription for the Windows Azure HDInsight Service with Hive running on it. These details were discussed previously Chapters 2 and 3.

Creating the Project

SQL Server Data Tools (SSDT) is the integrated development environment available from Microsoft to design, deploy, and develop SSIS packages. SSDT is installed when you choose to install SQL Server Client tools and Workstation Components from your SQL Server installation media. SSDT supports the creation of Integration Services, Analysis Services, and Reporting Services projects. Here, the focus is on the Integration Services project type.

To begin designing the package, load SQL Server Data Tools from the SQL Server 2012 program folders as in Figure 10-1.

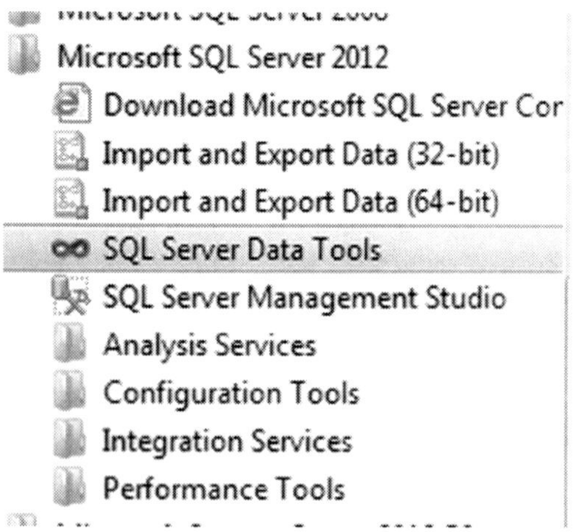

Figure 10-1. *SQL Server data tools*

Create a new project, and choose Integration Services Project in the New Project dialog as shown in Figure 10-2.

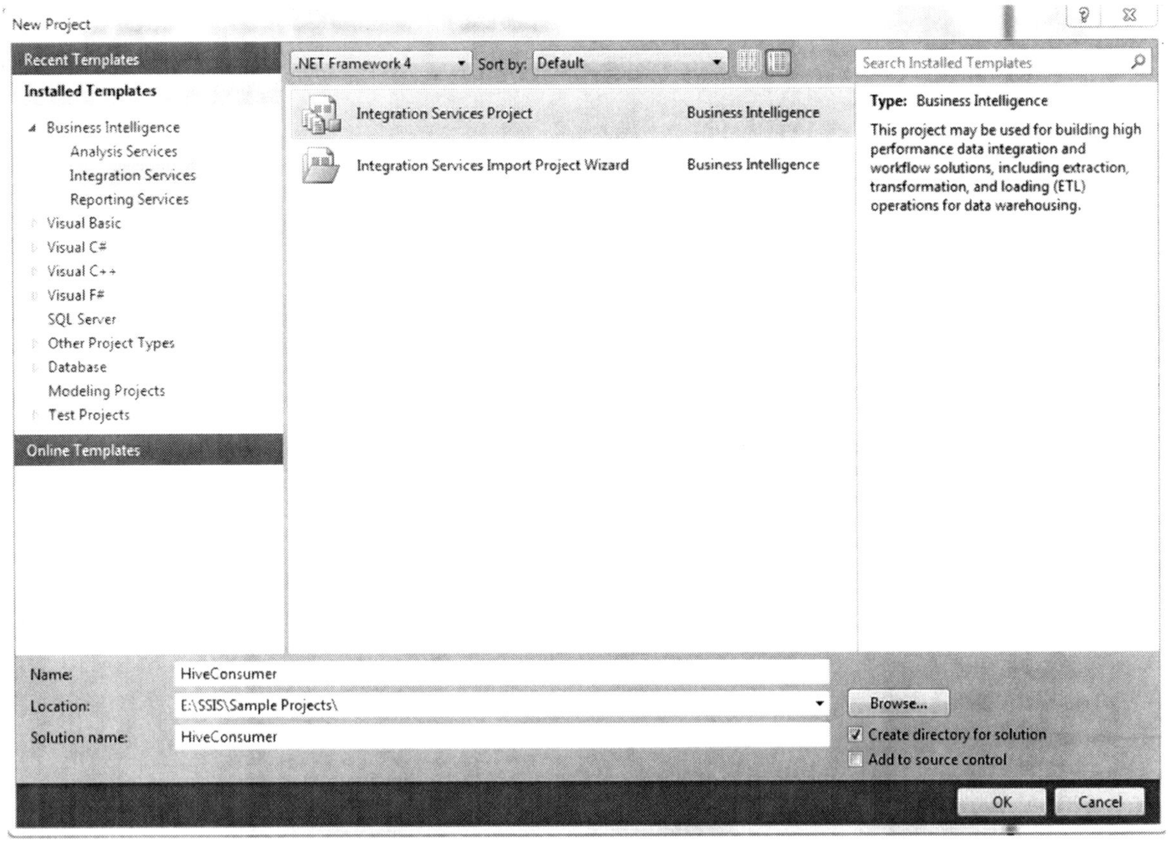

Figure 10-2. *New SSIS project*

When you select the Integration Services Project option, an SSIS project with a blank package named *Package. dtsx* is created. This package is visible in the Solution Explorer window of the project as shown in Figure 10-3.

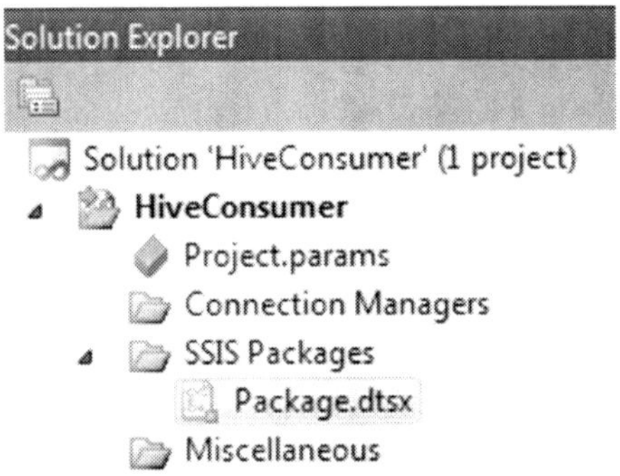

Figure 10-3. *Package.dtsx*

An SSIS solution is a placeholder for a meaningful grouping of different SSIS workflows. It can have multiple projects (in this solution, you have only one, *HiveConsumer*), and each project, in turn, can have multiple SSIS packages (in this project, you have only one, *Package.dtsx*) implementing specific data-load jobs.

Creating the Data Flow

As discussed earlier, a *data flow* is an SSIS package component used for moving data across different sources and destinations. In this package, to move data from Hive to SQL Server, you first need to create a data flow task in the package that contains the source and destination components to transfer the data.

Double Click the *Package.dtsx* created above in the SSDT solution to open the designer view. To create a data flow task, double-click (or drag and drop) a data flow task from the toolbox on the left side of the pane. This places a data flow task in the Control Flow canvas of the package as shown in Figure 10-4.

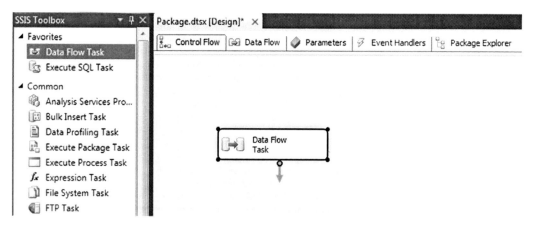

Figure 10-4. *SSIS data flow task*

Double-click the data flow task or click the Data Flow tab in SSDT to edit the task and design the source and destination components as shown in Figure 10-5.

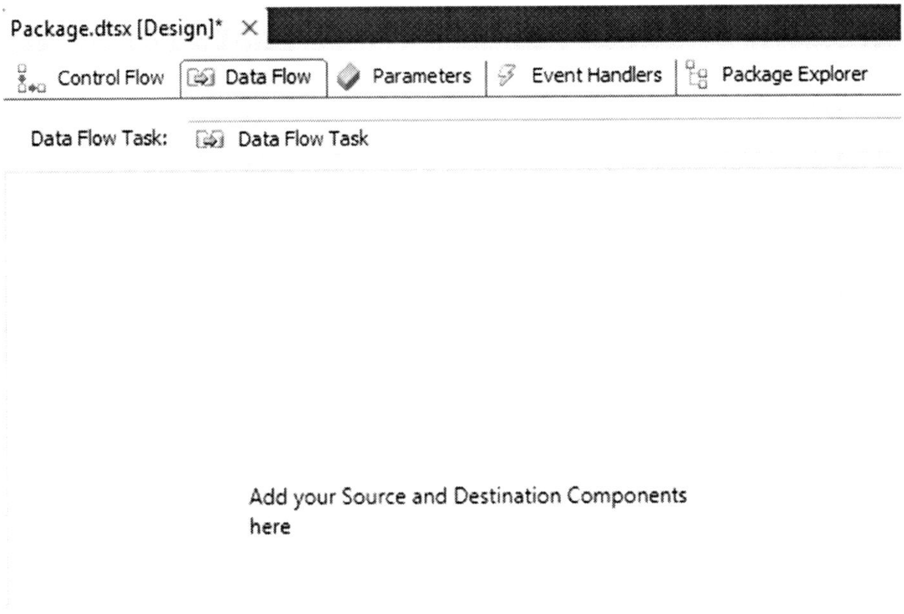

Package.dtsx [Design]* ✕

- Control Flow
- Data Flow
- Parameters
- Event Handlers
- Package Explorer

Data Flow Task: Data Flow Task

Add your Source and Destination Components here

Figure 10-5. The Data Flow tab

Creating the Source Hive Connection

Now it's time to create the connection to the Hive source. First create a connection manager that will connect to your Hive data tables hosted in HDInsight. You will use an ADO .NET connection, which will use the Data Source HadoopOnAzure you created on Chapter 8, to connect to Hive. To create the connection, right-click in the Connection Managers section in the project and select New ADO .Net Connection as shown in Figure 10-6.

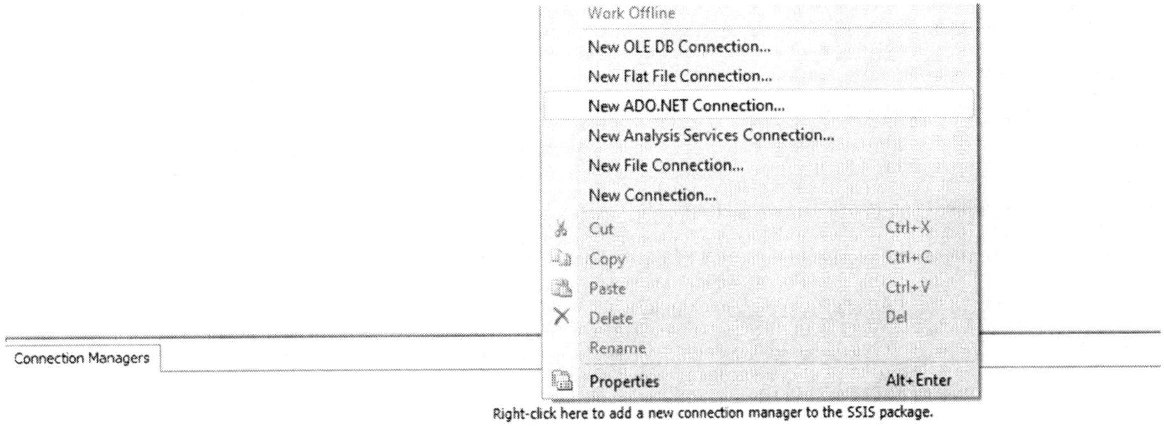

Work Offline		
New OLE DB Connection...		
New Flat File Connection...		
New ADO.NET Connection...		
New Analysis Services Connection...		
New File Connection...		
New Connection...		
✂ Cut	Ctrl+X
▣ Copy	Ctrl+C
▣ Paste	Ctrl+V
✕ Delete	Del
Rename	
▣ Properties	Alt+Enter

Connection Managers

Right-click here to add a new connection manager to the SSIS package.

Figure 10-6. Creating a new connection manager

Click the New button in the Configure Ado.Net Connection Manager window to create the new connection. From the list of providers, Select .Net Providers ➤ ODBC Data Provider and click OK in the Connection Manager window as shown in the Figure 10-7.

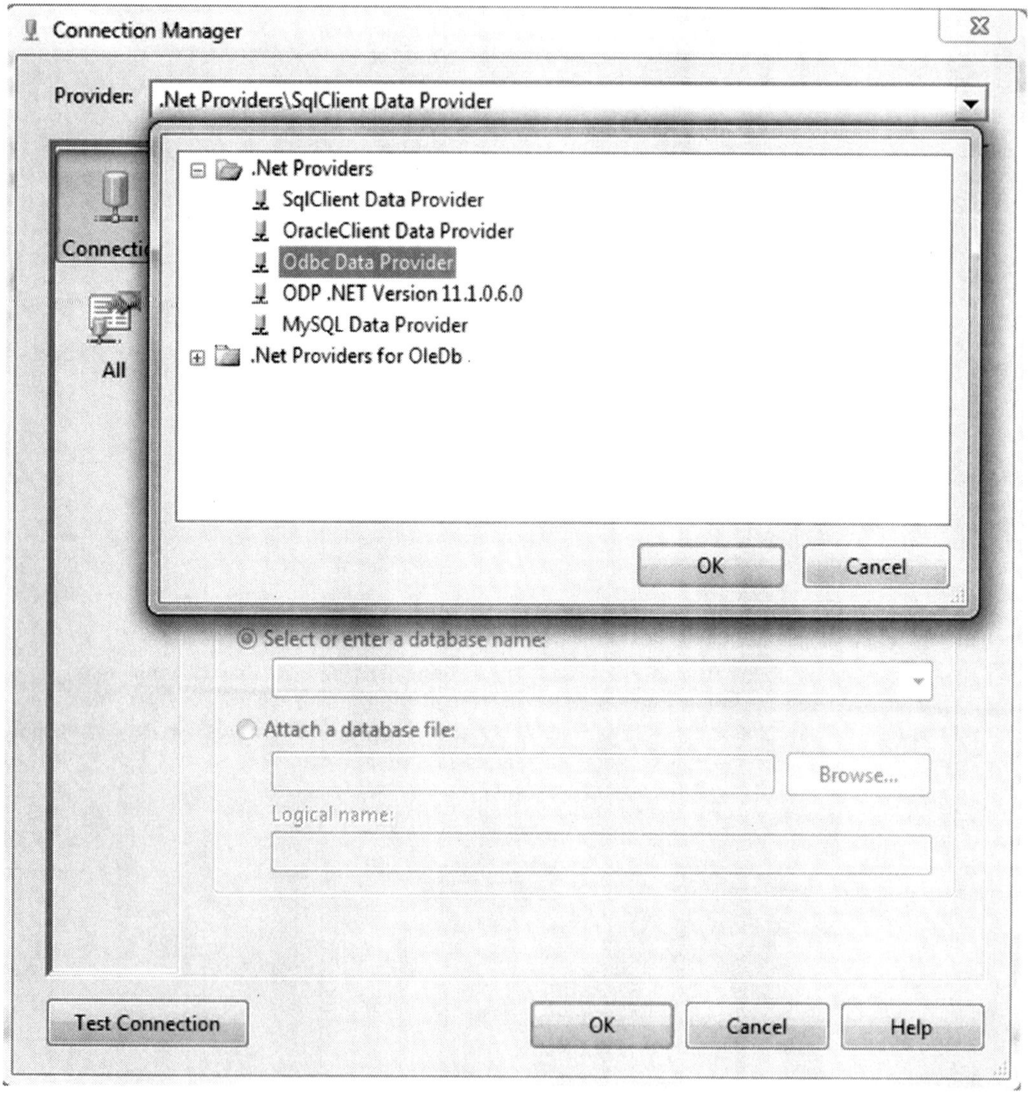

Figure 10-7. *Choosing the .NET ODBC Data Provider*

Select the HadoopOnAzure DSN from the User DSN or System DSN list, depending upon the type of DSN you created in Chapter 8. Provide the HDInsight cluster credentials and Test Connection should succeed as shown in Figure 10-8.

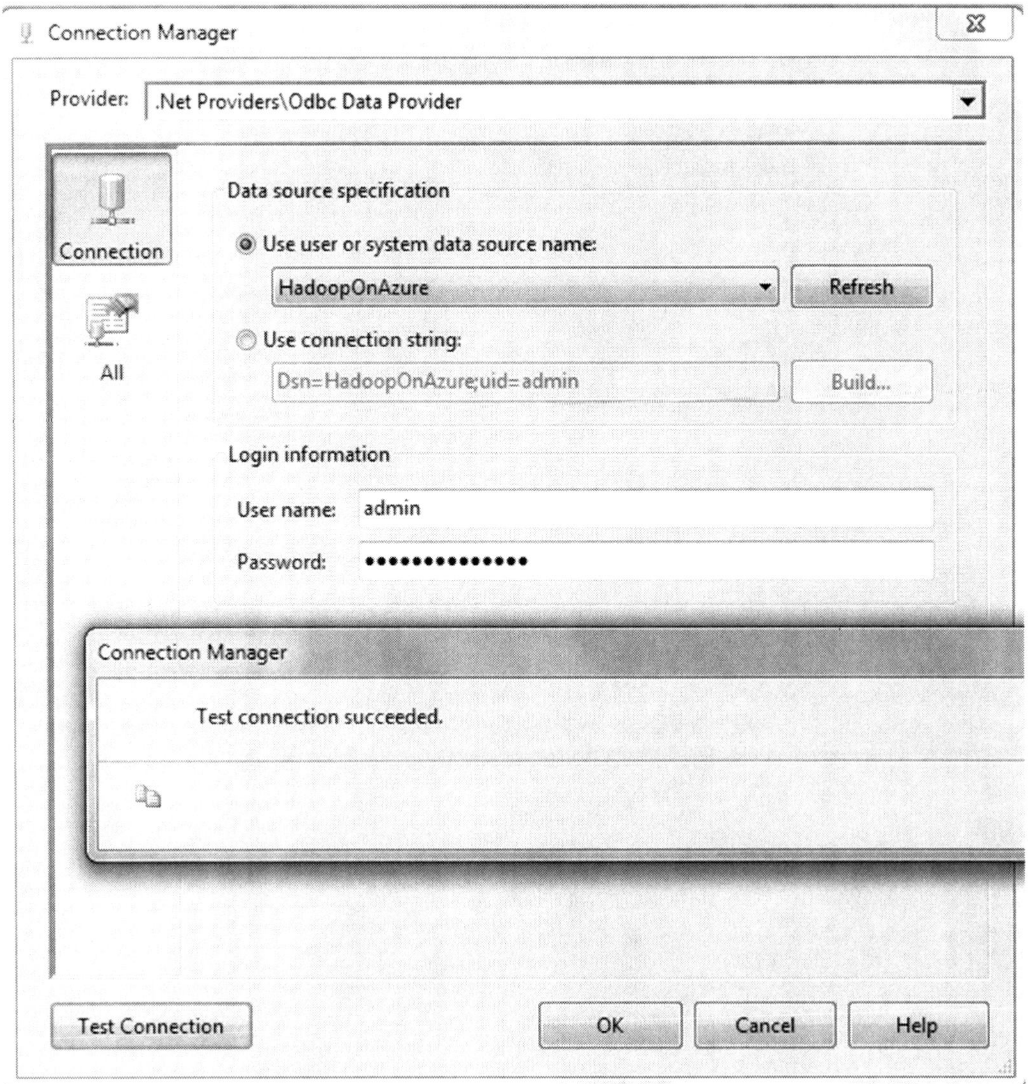

Figure 10-8. *Test the connection to Hive*

Creating the Destination SQL Connection

You need to configure a connection to point to the SQL Server instance and the database table where you will import data from Hive. For this, you need to create a connection manager to the destination SQL as you did for the source Hive. Right-click in the Connection Managers section of the project again, and this time, choose New OLE DB Connection as shown in Figure 10-9.

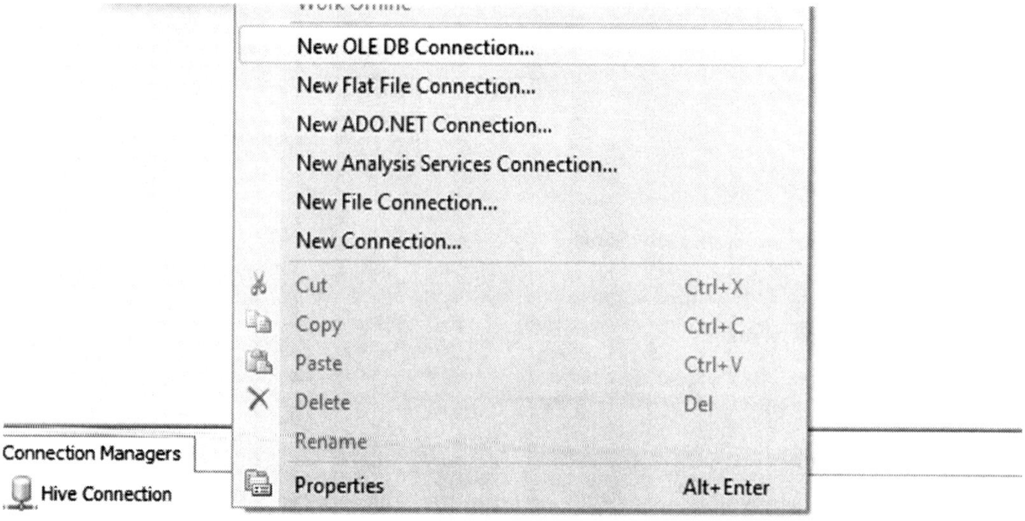

Figure 10-9. *Creating a new OLE DB connection to SQL Server*

From the list of providers, select Native OLE DB ➤ SQL Server Native Client 11.0. Type the name of the target SQL Server instance, and select the database where the target table resides. The test connection should succeed, thereby confirming the validity of the connection manager for the destination as shown in following Figure 10-10.

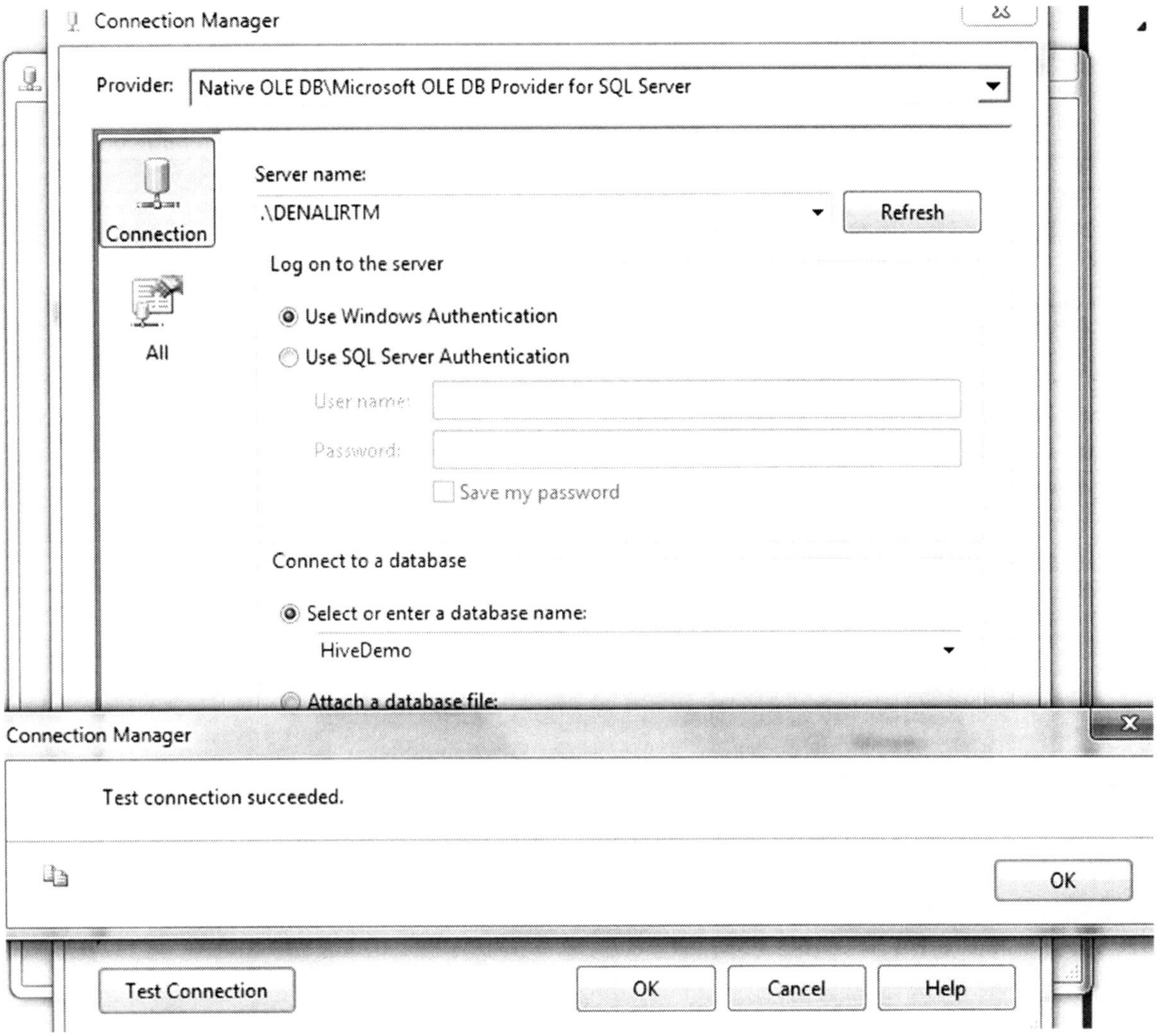

Figure 10-10. *Testing the SQL connection*

■ **Note** In this example, I chose OLE DB to connect to SQL. You can also choose to use ADO .NET or an ODBC connection to do the same. Also, a SQL database HiveDemo is pre-created using SQL Server Management Studio.

Creating the Hive Source Component

Next, you need to configure a source component that will connect to Hive and fetch the data. After the connection is successfully created, double-click to place an ADO NET source on the Data Flow canvas as shown in Figure 10-11.

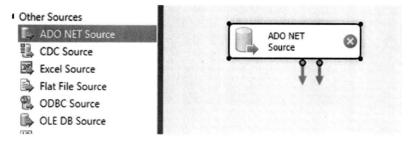

Figure 10-11. *Creating the ADO .NET source*

■ **Note** ODBC Source and ODBC Destination are a pair of data-flow components that included with SSIS 2012. The lack of direct SSIS ODBC components was always a complaint from customers regarding the product; hence, Microsoft partnered with Attunity to make these components available as a part of the product. Though the ODBC Source component supports many ODBC-compliant data sources, it does not currently support the Hive ODBC driver. Today, the only option to consume the Hive ODBC driver from SSIS is via the ADO.NET components.

Right-click the ADO.NET source and select *Edit* to configure the source to connect to the target Hive table using the connection you just created. Select the connection manager (I named it *Hive Connection*) and the Hive table (in this case, it's the *stock_analysis* table you created in Chapter 8) as shown in Figure 10-12.

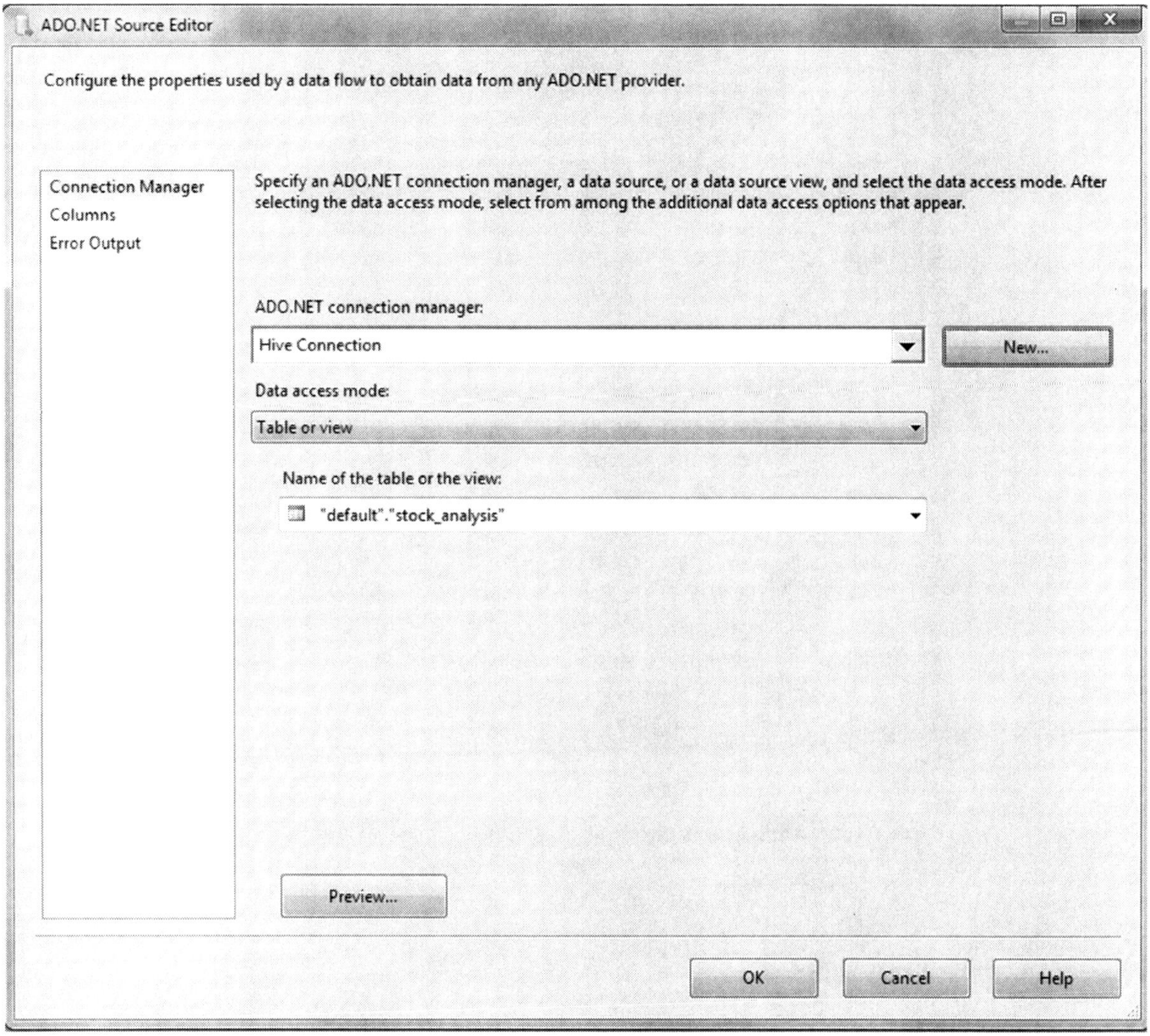

Figure 10-12. *Selecting the Hive table*

■ **Tip** You also can create the connection manager on the fly while configuring the source component by clicking the New button adjacent to the ADO.NET connection manager.

Click on the *Preview* button to preview the data, and ensure that it is being fetched from the source without issue. You should be able to see the first few rows from your Hive table, as in Figure 10-13.

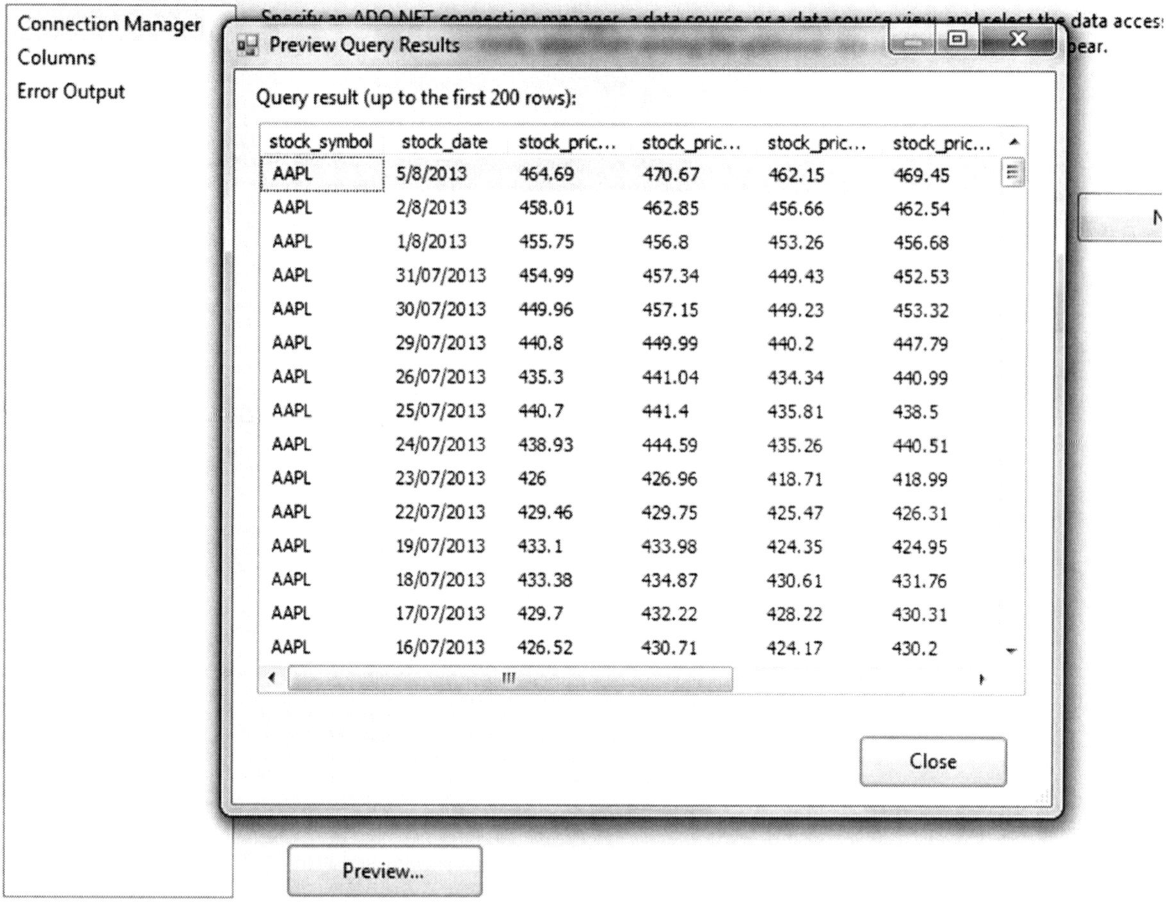

Connection Manager

Columns

Error Output

Figure 10-13. *Preview Hive query results*

Navigate to the *Columns* tab. Confirm that all columns from the source Hive table are detected and fetched as shown in Figure 10-14.

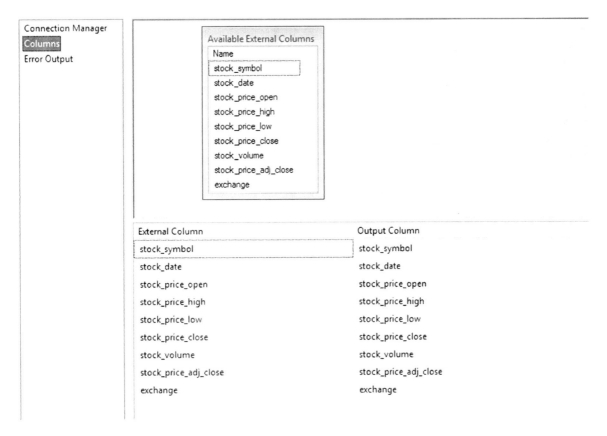

Figure 10-14. *Hive table columns*

Creating the SQL Destination Component

After the source is configured, you need to configure the destination where you want to import the Hive data. In this example, I use SQL Server as the destination. To do this, double-click on *OLE DB Destination* component in the Toolbox, and place an *OLE DB Destination* component on the *Data Flow* canvas. Make sure you connect the *ADO. NET source* and the *OLE DB Destination* components by dragging the arrow between the source and the destination. This is required for SSIS to generate the metadata and the column mappings for the destination automatically based on the source schema structure. The package should look something like Figure 10-15.

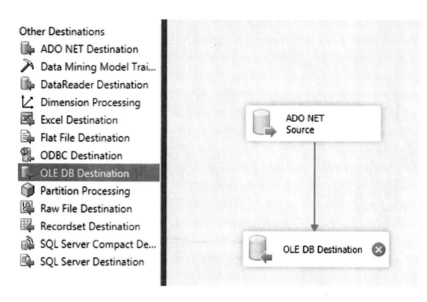

Figure 10-15. Creating the OLE DB destination

■ **Note** In this example, I used the OLE DB Destination component to bind to the target SQL Server table. However, you also can use the ADO .NET Destination or SQL Server Destination components for the same purpose. However, be aware that SQL Server Destination works only if the package runs locally on the same system where SQL Server resides.

Now, it is time to configure the *OLE DB Destination* component to point to the correct SQL connection and database table. To do this, right-click the *OLE DB Destination* component and select *Edit*. Select the OLE DB connection manager to SQL that you just created and the target table. In this case, I named the connection *SQL Connection* and predefined a table created in the SQL database called *stock_analysis*. If you don't have the table precreated, you can choose to create the destination table on the fly by clicking the *New* button adjacent to the name of the table or the view drop-down list. This is illustrated in Figure 10-16.

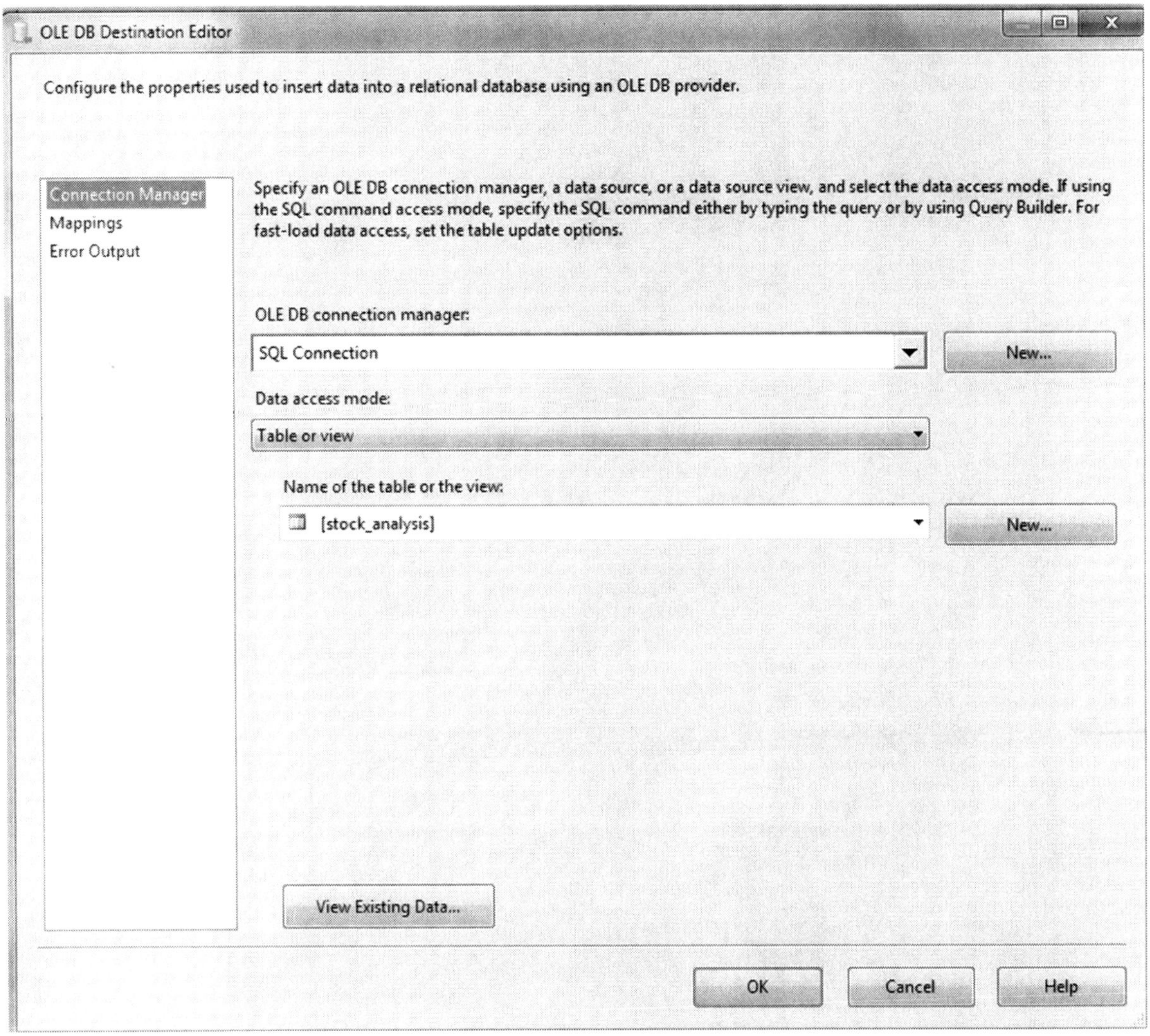

Figure 10-16. *Choosing the target SQL Server table*

■ **Note**   You also can create the connection manager and the database table on the fly while configuring the destination component by clicking on the respective *New* buttons as shown in Figure 10-16.

Mapping the Columns

After you set up the connection manager and select the destination table, navigate to the *Mappings* tab to ensure the column mappings between the source and the destination are correct, as shown in Figure 10-17. Click on *OK* to complete the configuration.

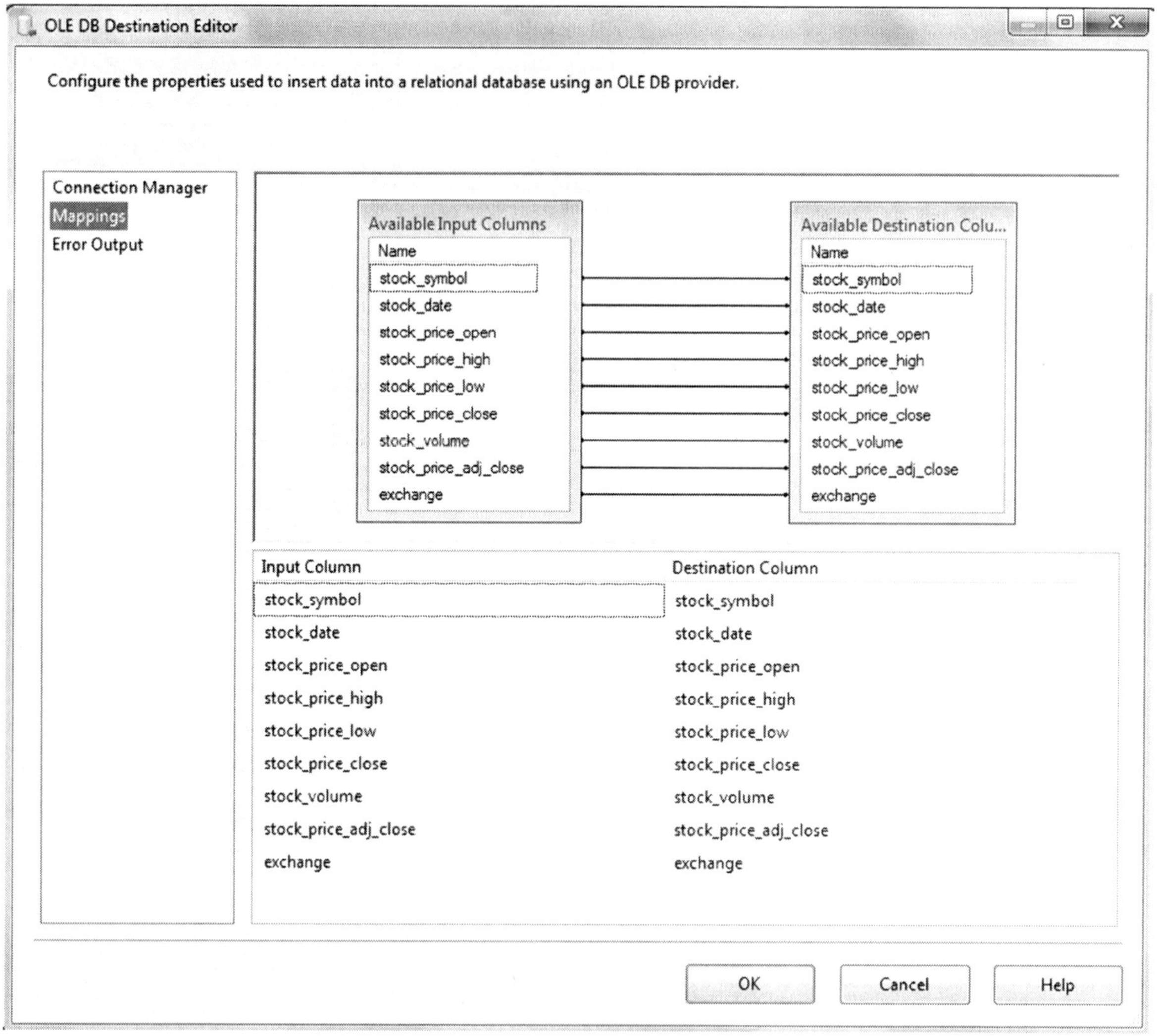

Figure 10-17. Verifying the column mappings

■ **Caution** If you choose to create the target table yourself and specify different column names than the source, you have to manually map each of these source and destination columns. SSIS's inbuilt column-mapping intelligence is based on having the same column names, so if they differ, make sure you set up the column mappings correctly.

The data flow with the source and destination, along with the connection managers, should like Figure 10-18.

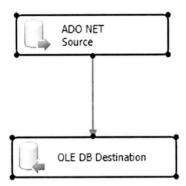

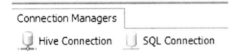

Figure 10-18. The complete data flow

Running the Package

Voila! You are all set to go. From the menu bar, select Debug ➤ Start Debugging, press *F5*, or click the Play button in the toolbar to execute the package as shown in Figure 10-19.

Figure 10-19. Executing the package

The package should run successfully, transfer records from the Hive table to the SQL Server table, and display the total number of records imported, as shown in Figure 10-20.

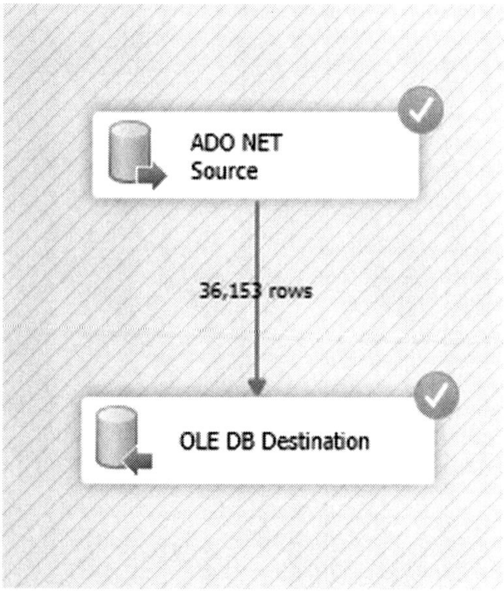

Figure 10-20. *Successful package execution*

If you are running this package in a 64-bit Windows operating system, you need to change the `Run64BitRuntime` property to `False`. This can be done from the Project Properties ➤ Configuration Properties ➤ Debugging tab to execute the package as it is using the 32-bit Hive ODBC driver, as shown in Figure 10-21.

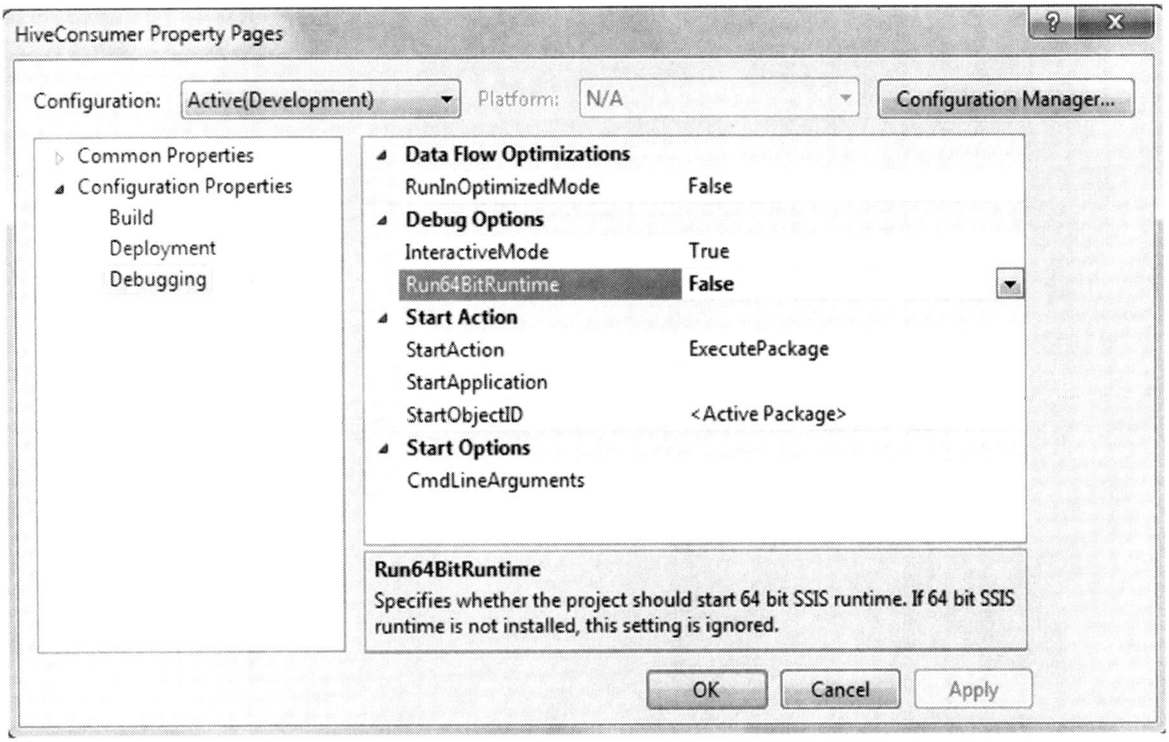

Figure 10-21. *Running in 32-bit mode*

You can now schedule this package as a SQL Server job and run the data load on a periodic basis. You also might want to apply some transformation to the data before it loads into the target SQL warehouse to clean it or to apply necessary business logic using the inbuilt SSIS Data Flow Transformation components.

There are other programmatic ways through which you can initiate a Hadoop job from SSIS. For example, you can develop your own custom SSIS components using .NET and use them to automate Hadoop jobs. A detailed description on this approach can be found on the following MSDN whitepaper:

```
http://msdn.microsoft.com/en-us/library/jj720569.aspx
```

Summary

In this chapter, you had a brief look into SQL Server and its Business Intelligence components. You also developed a sample package that connects to Hive using the Microsoft Hive ODBC Driver and imports data from the Hive table stock_analysis to SQL Server. Once the data is in SQL Server, you can leverage warehousing solutions like Analysis Services to slice and dice the data and use Reporting Services for powerful reporting on the data. This also enables you to integrate nonrelational data to be merged with traditional RDBMS data and extract information from it as a whole.

## CHAPTER 11

Logging in HDInsight

A complex eco system like Hadoop must have a detailed logging mechanism to fall back on in case something goes wrong. In traditional Hadoop, all the services—like NameNode, JobTracker, TaskTracker, and so on—have logging capabilities where each and every operation is logged right from service startup to shut down. Apart from the services or daemons startup, there are additional events that need to be recorded, such as job requests, interprocess communication between the services, job execution history, and others.

HDInsight distribution extends this logging mechanism by implementing its own. As you know, the entire cluster storage for the HDInsight service is in Azure in the form of blob containers. So you need to know and rely on the Azure storage logs to track down any access or space limitation issues. This chapter specifically focuses on the logging and instrumentation available for the Windows Azure-based Hadoop services and also gives you a glimpse into the traditional Hadoop logging mechanism.

Hadoop uses the Apache *Log4j* framework for logging, which is basically a logging package for Java. This logging framework not only logs operational information, it also gives you the control to tune different levels of logging as required—for example, errors or warnings—and several instrumentation options like log recycling, maintaining log history, and so on. This chapter will talk about a few key Log4j properties, but for a detailed understanding on the Log4j framework, you can visit the Apache site:

```
http://logging.apache.org/log4j/2.x/manual/index.html
```

Service Logs

Hadoop daemons are replaced by Windows Services in the HDInsight distribution. Different services run on different nodes of the cluster based on the role they play. You need to make a remote desktop connection to the nodes to access their respective log files.

Service Trace Logs

The service startup logs are located in the `C:\apps\dist\hadoop-1.2.0.1.3.1.0-06\bin` directory for the Hadoop services. Similarly, other service-based projects in the ecosystem (like Hive, Oozie and so on) log their service startup operations in their respective bin folders. These files are marked with `.trace.log` extensions, and they are created and written to during the startup of the services. Table 11-1 summarizes the different types of trace.log files available for the projects shipped in the current distribution of HDInsight on Azure.

Table 11-1. *Log files available in HDInsight on Azure*

Log File Name	Location	Service	Node
namenode.trace.log	C:\apps\dist\hadoop-1.2.0.1.3.1.0-06\bin	Hadoop Name Node Service	Cluster name node
datanode.trace.log	C:\apps\dist\hadoop-1.2.0.1.3.1.0-06\bin	Hadoop Data Node Service	Any of the cluster data nodes
secondarynamenode.trace.log	C:\apps\dist\hadoop-1.2.0.1.3.1.0-06\bin	Hadoop Secondary Name Node Service	Cluster secondary name node
tasktracker.trace.log	C:\apps\dist\hadoop-1.2.0.1.3.1.0-06\bin	Hadoop taskTracker Service	Any of the cluster data nodes.
hiveserver.trace.log	C:\apps\dist\hive-0.11.0.1.3.1.0-06\bin	Hive Thrift Service	Cluster node running Hive
hiveserver2.trace.log	C:\apps\dist\hive-0.11.0.1.3.1.0-06\bin	Hive Server 2 (with concurrent connection support)	Cluster node running Hive
metastore.trace.log	C:\apps\dist\hive-0.11.0.1.3.1.0-06\bin	Hive Meta Store Service (Hive storage)	Cluster node running Hive
Derbyserver.trace.log	C:\apps\dist\hive-0.11.0.1.3.1.0-06\bin	Hive Derby Server Service (Hive native storage)	Cluster node running Hive
oozieservice.out.log	C:\apps\dist\oozie-3.3.2.1.3.1.0-06\Service	Oozie Service	Cluster node running Oozie
Templeton.trace.log	C:\apps\dist\hcatalog-0.11.0.1.3.1.0-06\bin	Templeton Service	Cluster node running Templeton

Figure 11-1 will help you correlate the services to the startup logs that are listed in Table 11-1. The log file names are similar to the corresponding Windows service names that you see in the figure. For example, the Apache Hadoop NameNode service will log its operations to the namenode.trace.log file, and so on.

Name ▲
- Apache Hadoop Derbyserver
- Apache Hadoop hiveserver
- Apache Hadoop hiveserver2
- Apache Hadoop isotopejs
- Apache Hadoop jobtracker
- Apache Hadoop metastore
- Apache Hadoop namenode
- Apache Hadoop oozieservice
- Apache Hadoop templeton

Figure 11-1. *HDInsight services*

These logs record the messages and failures during service startup, if there are any. They also record the ID number of the process spawned when a service starts. Following is a sample namenode.trace.log file. It shows the content after a name node service startup.

```
HadoopServiceTraceSource Information: 0 : Tracing successfully initialized
        DateTime=2013-12-10T02:46:57.6055000Z
        Timestamp=3981555628
HadoopServiceTraceSource Information: 0 : Loading service xml: c:\apps\dist\hadoop-1.2.0.1.3.1.0-06\
bin\namenode.xml
        DateTime=2013-12-10T02:46:57.6055000Z
        Timestamp=3981598144
HadoopServiceTraceSource Information: 0 : Successfully parsed service xml for service namenode
        DateTime=2013-12-10T02:46:57.6211250Z
        Timestamp=3981610465
HadoopServiceTraceSource Information: 0 : Command line: c:\apps\dist\java\bin\java -server
-Xmx4096m  -Dhadoop.log.dir=c:\apps\dist\hadoop-1.2.0.1.3.1.0-06\logs -Dhadoop.log.file=hadoop-
namenode-RD00155D67172B.log -Dhadoop.home.dir=c:\apps\dist\hadoop-1.2.0.1.3.1.0-06 -Dhadoop.root.
logger=INFO,console,DRFA,ETW,FilterLog -Djava.library.path=;c:\apps\dist\hadoop-1.2.0.1.3.1.0-06\
lib\native\Windows_NT-amd64-64;c:\apps\dist\hadoop-1.2.0.1.3.1.0-06\lib\native -Dhadoop.
policy.file=hadoop-policy.xml -Dcom.sun.management.jmxremote  -Detwlogger.component=namenode
-Dwhitelist.filename=core-whitelist.res  -classpath c:\apps\dist\hadoop-1.2.0.1.3.1.0-06\
conf;c:\apps\dist\java\lib\tools.jar;c:\apps\dist\hadoop-1.2.0.1.3.1.0-06;c:\apps\dist\
hadoop-1.2.0.1.3.1.0-06\hadoop-ant-1.2.0.1.3.1.0-06.jar;c:\apps\dist\hadoop-1.2.0.1.3.1.0-06\hadoop-
client-1.2.0.1.3.1.0-06.jar;c:\apps\dist\hadoop-1.2.0.1.3.1.0-06\hadoop-core-1.2.0.1.3.1.0-06.
jar;c:\apps\dist\hadoop-1.2.0.1.3.1.0-06\hadoop-core.jar;c:\apps\dist\hadoop-1.2.0.1.3.1.0-06\
hadoop-examples-1.2.0.1.3.1.0-06.jar;c:\apps\dist\hadoop-1.2.0.1.3.1.0-06\hadoop-examples.
jar;c:\apps\dist\hadoop-1.2.0.1.3.1.0-06\hadoop-minicluster-1.2.0.1.3.1.0-06.jar;c:\apps\dist\
hadoop-1.2.0.1.3.1.0-06\hadoop-test-1.2.0.1.3.1.0-06.jar;c:\apps\dist\hadoop-1.2.0.1.3.1.0-06\
hadoop-test.jar;c:\apps\dist\hadoop-1.2.0.1.3.1.0-06\hadoop-tools-1.2.0.1.3.1.0-06.jar;c:\apps\dist\
hadoop-1.2.0.1.3.1.0-06\hadoop-tools.jar;c:\apps\dist\hadoop-1.2.0.1.3.1.0-06\lib\*;c:\apps\dist\
hadoop-1.2.0.1.3.1.0-06\lib\jsp-2.1\*;c:\apps\dist\log4jetwappender\microsoft-log4j-etwappender-
1.0.jar; org.apache.hadoop.hdfs.server.namenode.NameNode
        DateTime=2013-12-10T02:46:57.6211250Z
        Timestamp=3981611043
HadoopServiceTraceSource Information: 0 : ServiceHost#OnStart
        DateTime=2013-12-10T02:46:57.6211250Z
        Timestamp=3981662789
HadoopServiceTraceSource Information: 0 : Child process started, PID: 3720
        DateTime=2013-12-10T02:46:57.6211250Z
        Timestamp=3981707399
```

These logs record very low-level service startup messages. Most likely, the information in them is external to the Hadoop system. For example, in a network failure scenario, you might see an entry similar to the following in your namenode.trace.log file:

```
Session Terminated, Killing shell....
```

It is very rare that these log files get populated with anything else apart from the service startup messages. For example, they might be populated in the case of a network heartbeat failure between the name node and the data nodes. Still, they can be helpful at times in figuring out why your DataNode, NameNode, or Secondary NameNode service isn't starting up or is sporadically shutting down.

▪ **Note** These .trace.log files are introduced with HDInsight cluster version 2.1. In version 1.6 clusters, the file names are out.log.

The following two sections are specific to HDInsight clusters in version 1.6. The log file types discussed are not available if the cluster version is 2.1. This holds good for the Windows Azure HDInsight Emulator since, as of this writing, it deploys HDInsight components version 1.6. In all probability, the HDInsight emulator will be upgraded soon to match the version of the Azure service and both will have same set of log files.

Service Wrapper Files

Apart from the startup logs, there are something called wrapper logs available for the HDInsight services. These files contain the startup command string to start the service. It also provides the output of the process id when the service starts successfully. They are of .wrapper.log extension and are available in the same directory where the .out.log files reside. For example, if you open hiveserver.wrapper.log you should see commands similar to the snippet below.

```
org.apache.hadoop.hive.service.HiveServer -hiveconf hive.hadoop.classpath=c:\apps\dist\hive-0.9.0\
lib\* -hiveconf hive.metastore.local=true -hiveconf hive.server.servermode=http -p 10000  -hiveconf
hive.querylog.location=c:\apps\dist\hive-0.9.0\logs\history -hiveconf hive.log.dir=c:\apps\dist\
hive-0.9.0\logs
2013-08-11 16:40:45 - Started 4264
```

Note that the process id of the service is recorded at the end of the wrapper log. This is very helpful in troubleshooting scenarios where you may want to trace on a specific process which has already started, for example, determining the heap memory usage of the name node process when it is running while troubleshooting an out of memory problem.

Service Error Files

The HDInsight version 1.6 services also generate an error log file for each service. These record the log messages for the running java services. If there are any errors encountered while the service is already running, the stack trace of the error is logged in the above files. The error logs are of extension .err.log and they again, reside on the same directory as the output and wrapper files. For example, if you have permission issues in accessing the required files and folders, you may see an error message similar to below in your namenode.err.log file.

```
13/08/16 19:07:16 WARN impl.MetricsSystemImpl: Source name ugi already exists!
13/08/16 19:07:16 INFO util.GSet: VM type        = 64-bit
13/08/16 19:07:16 INFO util.GSet: 2% max memory = 72.81875 MB
13/08/16 19:07:16 INFO util.GSet: capacity       = 2^23 = 8388608 entries
13/08/16 19:07:16 INFO util.GSet: recommended=8388608, actual=8388608
13/08/16 19:07:16 INFO namenode.FSNamesystem: fsOwner=admin
13/08/16 19:07:16 INFO namenode.FSNamesystem: supergroup=supergroup
13/08/16 19:07:16 INFO namenode.FSNamesystem: isPermissionEnabled=false
13/08/16 19:07:16 INFO namenode.FSNamesystem: dfs.block.invalidate.limit=100
13/08/16 19:07:16 ERROR namenode.FSNamesystem: FSNamesystem initialization failed.
java.io.FileNotFoundException: c:\hdfs\nn\current\VERSION (Access is denied)
        at java.io.RandomAccessFile.open(Native Method)
        at java.io.RandomAccessFile.<init>(RandomAccessFile.java:233)
        at org.apache.hadoop.hdfs.server.common.Storage$StorageDirectory.read(St
orage.java:222)
```

I am running all the services for my demo cluster in the name node itself. My set of Hadoop service log files for cluster version 2.1 looks like those shown in Figure 11-2.

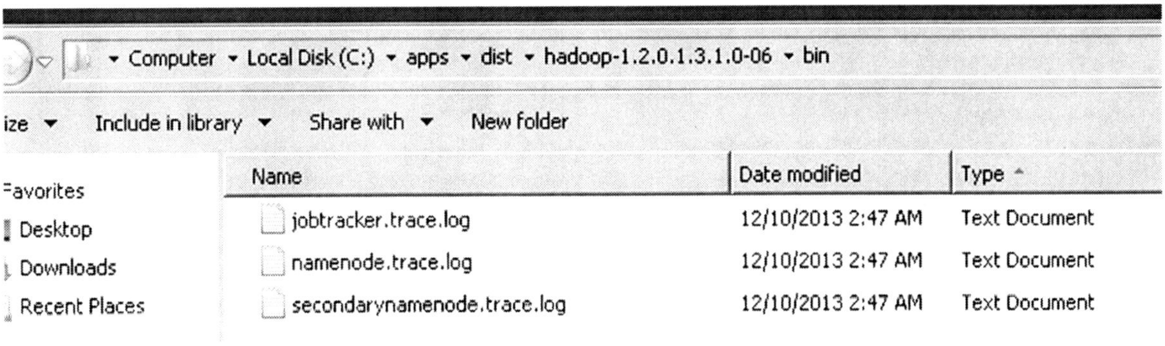

Figure 11-2. *Hadoop service log files*

The service log files are common for all the services listed in Table 11-1. That means that each of the service-based projects, like Hive and so on, have these sets of service log files in their respective bin folders.

Hadoop log4j Log Files

When you consider that HDInsight is essentially a wrapper on top of core Hadoop; it is no surprise that it continues to embrace and support the traditional logging mechanism by Apache. You should continue to investigate these log files for most of your job failures, authentication issues, and service communication issues.

In the HDInsight distribution on Azure, these logs are available in the C:\apps\dist\hadoop-1.2.0.1.3.1.0-06\ logs directory of the respective nodes for Hadoop. By default, the log files are recycled daily at midnight; however, historical versions are preserved. The old file names are appended with a _<*timestamp*> value each time they are purged and rolled over. The most current log files are in the format *hadoop-namenode-<Hostname>.log,* *hadoop-datanode-<Hostname>.log, hadoop-secondarynamenode-<Hostname>.log,* and so on. The *Hostname* is the host where the service is running on. These are pretty similar to the service error-log files discussed in the previous section and record the stack traces of the service failures. A typical name node log looks similar to the following snippet after a successful startup.

```
2013-08-16 21:32:39,324 INFO org.apache.hadoop.hdfs.server.namenode.NameNode: STARTUP_MSG:
/************************************************************
STARTUP_MSG: Starting NameNode
STARTUP_MSG:   host = <HostName>/<IP Address>
STARTUP_MSG:   args = []
STARTUP_MSG:   version = 1.2.0
STARTUP_MSG:   build = git@github.com:hortonworks/hadoop-monarch.git on branch (no branch)
 -r 99a88d4851ce171cf57fa621910bb293950e6358; compiled by 'jenkins' on Fri Jul 19 22:07:17
 Coordinated Universal Time 2013
************************************************************/
```

```
2013-08-16 21:32:40,167 WARN org.apache.hadoop.metrics2.impl.MetricsSystemImpl:
    Source name ugi already exists!
2013-08-16 21:32:40,199 INFO org.apache.hadoop.hdfs.util.GSet: VM type        = 64-bit
2013-08-16 21:32:40,199 INFO org.apache.hadoop.hdfs.util.GSet: 2% max memory = 72.81875 MB
2013-08-16 21:32:40,199 INFO org.apache.hadoop.hdfs.util.GSet: capacity       = 2^23 =
    8388608 entries
2013-08-16 21:32:40,199 INFO org.apache.hadoop.hdfs.util.GSet: recommended=8388608,
    actual=8388608
2013-08-16 21:32:40,245 INFO org.apache.hadoop.hdfs.server.namenode.FSNamesystem: fsOwner=hdp
2013-08-16 21:32:40,245 INFO org.apache.hadoop.hdfs.server.namenode.FSNamesystem:
    supergroup=supergroup
2013-08-16 21:32:40,245 INFO org.apache.hadoop.hdfs.server.namenode.FSNamesystem:
    isPermissionEnabled=false
2013-08-16 21:32:40,261 INFO org.apache.hadoop.hdfs.server.namenode.FSNamesystem:
    dfs.block.invalidate.limit=100
2013-08-16 21:32:40,261 INFO org.apache.hadoop.hdfs.server.namenode.FSNamesystem:
    isAccessTokenEnabled=false accessKeyUpdateInterval=0 min(s), accessTokenLifetime=0 min(s)
2013-08-16 21:32:40,292 INFO org.apache.hadoop.hdfs.server.namenode.FSNamesystem: Registered
    FSNamesystemStateMBean and NameNodeMXBean
2013-08-16 21:32:40,355 INFO org.apache.hadoop.hdfs.server.namenode.FSEditLog:
    dfs.namenode.edits.toleration.length = 0
2013-08-16 21:32:40,355 INFO org.apache.hadoop.hdfs.server.namenode.NameNode:
    Caching file names occuring more than 10 times
2013-08-16 21:32:40,386 INFO org.apache.hadoop.hdfs.server.namenode.FSEditLog:
    Read length      = 4
2013-08-16 21:32:40,386 INFO org.apache.hadoop.hdfs.server.namenode.FSEditLog:
    Corruption length = 0
2013-08-16 21:32:40,386 INFO org.apache.hadoop.hdfs.server.namenode.FSEditLog:
    Toleration length = 0 (= dfs.namenode.edits.toleration.length)
2013-08-16 21:32:40,386 INFO org.apache.hadoop.hdfs.server.namenode.FSEditLog: Summary:
    |---------- Read=4 ----------|-- Corrupt=0 --|-- Pad=0 --|
2013-08-16 21:32:41,855 INFO org.apache.hadoop.http.HttpServer: Port returned by
webServer.getConnectors()[0].getLocalPort() before open() is -1. Opening the listener on 50070
2013-08-16 21:32:41,855 INFO org.apache.hadoop.http.HttpServer: listener.getLocalPort()
    returned 50070 webServer.getConnectors()[0].getLocalPort() returned 50070
2013-08-16 21:32:41,855 INFO org.apache.hadoop.http.HttpServer: Jetty bound to port 50070
2013-08-16 21:32:42,527 INFO org.apache.hadoop.hdfs.server.namenode.NameNode:
    Web-server up at: namenodehost:50070
2013-08-16 21:32:42,558 INFO org.apache.hadoop.ipc.Server: IPC Server listener on 9000: starting
2013-08-16 21:32:42,574 INFO org.apache.hadoop.ipc.Server: IPC Server handler 1 on 9000: starting
2013-08-16 21:32:42,574 INFO org.apache.hadoop.ipc.Server: IPC Server handler 7 on 9000: starting
2013-08-16 21:32:42,574 INFO org.apache.hadoop.ipc.Server: IPC Server handler 5 on 9000: starting
```

The log gives you important information like the host name, the port number on which the web interfaces listen, and a lot of other storage-related information that could be useful while troubleshooting a problem. In the case of an authentication problem with the data nodes, you might see error messages similar to the following one in the logs:

```
2013-08-16 21:32:43,152 ERROR org.apache.hadoop.security.UserGroupInformation:
PriviledgedActionException as:hdp cause:java.io.IOException: File /mapred/system/jobtracker.info
could only be replicated to 0 nodes, instead of 1.
```

> **Note** Each message in the logs is marked by levels like INFO, ERROR, and so on. This level of verbosity in the error logs can be controlled using the Log4j framework.

Figure 11-3 shows a screenshot of the Hadoop log files for my democluster.

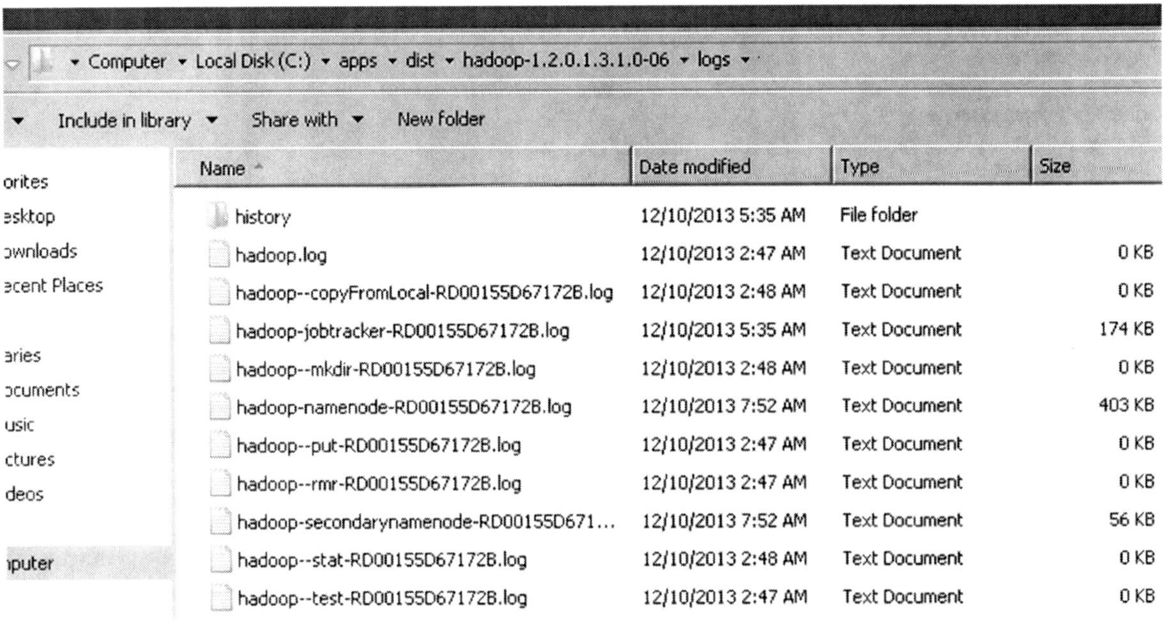

Figure 11-3. *Hadoop Log4j logs*

A few of the supporting projects like Hive also support the Log4j framework. They have these logs in their own log directory similar to Hadoop. Following is a snippet of my Hive server log files running on democluster.

```
(HiveMetaStore.java:main(2940)) - Starting hive metastore on port 9083
2013-08-16 21:24:32,437 INFO  metastore.HiveMetaStore (HiveMetaStore.java:newRawStore(349)) - 0:
Opening raw store with implemenation class:org.apache.hadoop.hive.metastore.ObjectStore
2013-08-16 21:24:32,469 INFO  mortbay.log (Slf4jLog.java:info(67)) - Logging to
org.slf4j.impl.Log4jLoggerAdapter(org.mortbay.log) via org.mortbay.log.Slf4jLog
2013-08-16 21:24:32,515 INFO  metastore.ObjectStore (ObjectStore.java:initialize(206))
- ObjectStore, initialize called
2013-08-16 21:24:32,578 INFO  metastore.HiveMetaStore (HiveMetaStore.java:newRawStore(349)) - 0:
Opening raw store with implemenation class:org.apache.hadoop.hive.metastore.ObjectStore
2013-08-16 21:24:32,625 INFO  metastore.ObjectStore (ObjectStore.java:initialize(206))
- ObjectStore, initialize called
(HiveMetaStore.java:startMetaStore(3032)) - Starting DB backed MetaStore Server
2013-08-16 21:24:40,090 INFO  metastore.HiveMetaStore (HiveMetaStore.java:startMetaStore(3044))
- Started the new metaserver on port [9083]…
2013-08-16 21:24:40,090 INFO  metastore.HiveMetaStore (HiveMetaStore.java:startMetaStore(3046))
```

```
- Options.minWorkerThreads = 200
2013-08-16 21:24:40,090 INFO  metastore.HiveMetaStore (HiveMetaStore.java:startMetaStore(3048))
- Options.maxWorkerThreads = 100000
2013-08-16 21:24:40,091 INFO  metastore.HiveMetaStore (HiveMetaStore.java:startMetaStore(3050))
- TCP keepalive = true
2013-08-16 21:24:40,104 INFO  metastore.HiveMetaStore (HiveMetaStore.java:logInfo(392))
- 1: get_databases: default
2013-08-16 21:24:40,123 INFO  metastore.HiveMetaStore
Logging initialized using configuration in file:/C:/apps/dist/hive-0.9.0/conf/hive-
log4j.properties
2013-08-16 21:25:03,078 INFO  ql.Driver (PerfLogger.java:PerfLogBegin(99)) - <PERFLOG
 method=Driver.run>
2013-08-16 21:25:03,078 INFO  ql.Driver (PerfLogger.java:PerfLogBegin(99)) - <PERFLOG
 method=compile>
2013-08-16 21:25:03,145 INFO  parse.ParseDriver (ParseDriver.java:parse(427))
- Parsing command: DROP TABLE IF EXISTS HiveSampleTable
2013-08-16 21:25:03,445 INFO  parse.ParseDriver (ParseDriver.java:parse(444))
- Parse Completed
2013-08-16 21:25:03,541 INFO  hive.metastore (HiveMetaStoreClient.java:open(195))
- Trying to connect to metastore with URI thrift://headnodehost:9083
2013-08-16 21:25:03,582 INFO  hive.metastore (HiveMetaStoreClient.java:open(209))
- Connected to metastore.
2013-08-16 21:25:03,604 INFO  metastore.HiveMetaStore (HiveMetaStore.java:logInfo(392))
- 4: get_table : db=default tbl=HiveSampleTable
```

Again, the preceding log output is stripped for brevity, but you can see how the log emits useful information, such as several port numbers, the query that it fires to load the default tables, the number of worker threads, and much more. In the case of a Hive processing error, this log is the best place to look for further insight into the problem.

■ **Note**   A lot of documentation is available on Apache's site regarding the logging framework that Hadoop and its supporting projects implement. That information is not covered in depth in this chapter, which focuses on HDInsight-specific features.

Log4j Framework

There are a few key properties in the Log4j framework that will help you maintain your cluster storage more efficiently. If all the services are left with logging every bit of detail in the log files, a busy Hadoop cluster can easily run you out of storage space, especially in scenarios where your name node runs most of the other services as well. Such logging configurations can be controlled using the Log4j.properties file present in the conf directory for the projects. For example, Figure 11-4 shows the configuration file for my Hadoop cluster.

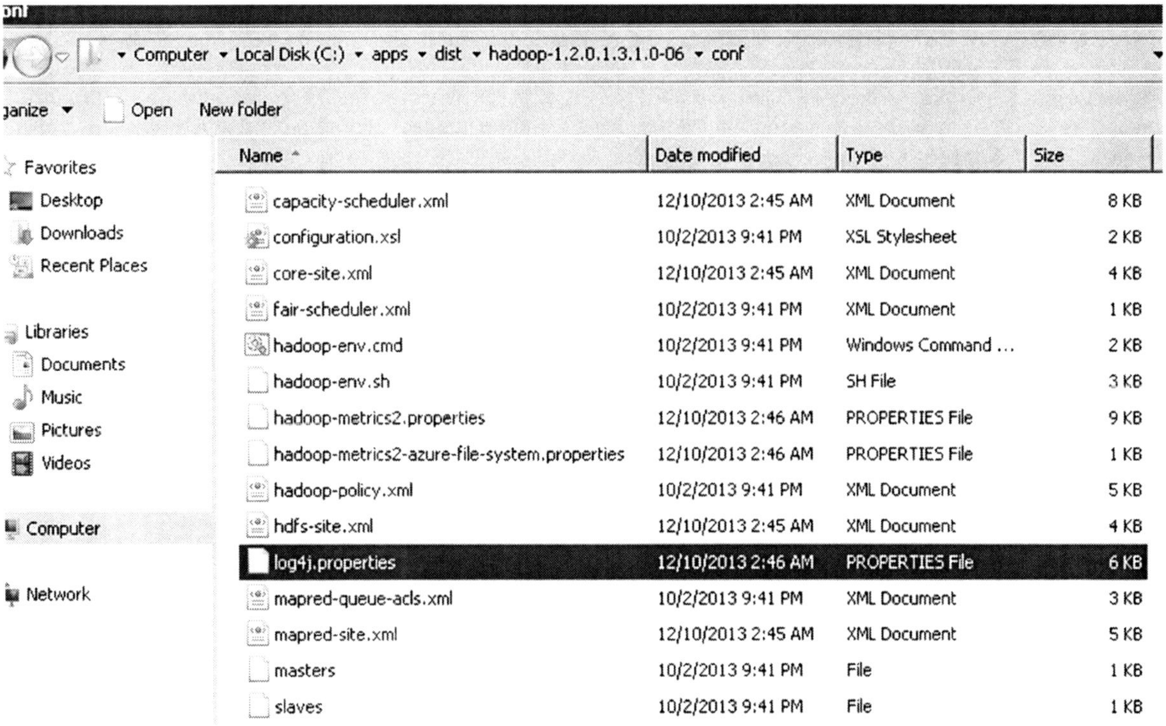

Figure 11-4. *Log4j.properties file*

There is a section in the file where you can specify the level of details to be recorded. The following code shows a snippet of the properties file:

```
#
# FSNamesystem Audit logging
# All audit events are logged at INFO level
#
log4j.logger.org.apache.hadoop.hdfs.server.namenode.FSNamesystem.audit=WARN

# Custom Logging levels

hadoop.metrics.log.level=WARN
#log4j.logger.org.apache.hadoop.mapred.JobTracker=DEBUG
#log4j.logger.org.apache.hadoop.mapred.TaskTracker=DEBUG
#log4j.logger.org.apache.hadoop.fs.FSNamesystem=DEBUG
log4j.logger.org.apache.hadoop.metrics2=${hadoop.metrics.log.level}

# Set the warning level to WARN to avoid having info messages leak
# to the console
log4j.logger.org.mortbay.log=WARN
```

The file is commented to make it easier for you to set the logging levels. As you can see in the preceding code example, you can set the log levels to WARN to stop logging generic INFO messages. You can opt to log messages only in the case of debugging for several services, like JobTracker and TaskTracker. To further shrink the logs, you can also set the logging level to ERR to ignore all warnings and worry only in the case of errors. There are other properties of interest as well, especially those that control the log rollover, retention period, maximum file size, and so on, as shown in the following snippet:

```
# Roll over at midnight
log4j.appender.DRFA.DatePattern=.yyyy-MM-dd

# 30-day backup
#log4j.appender.DRFA.MaxBackupIndex=30
log4j.appender.DRFA.layout=org.apache.log4j.PatternLayout
#Default values
hadoop.tasklog.taskid=null
hadoop.tasklog.iscleanup=false
hadoop.tasklog.noKeepSplits=4
hadoop.tasklog.totalLogFileSize=100
hadoop.tasklog.purgeLogSplits=true
hadoop.tasklog.logsRetainHours=12
```

Simple settings like these can really help you control log file growth and avoid certain problems in the future. You have limited control over what your users decide to log in their MapReduce code, but what you do have control over is the task attempt and execution log levels.

Each of the data nodes have a userlogs folder inside the C:\apps\dist\hadoop-1.2.0.1.3.1.0-06\logs\ directory. This folder contains a historical record of all the MapReduce jobs or tasks executed in the cluster. To create a complete chain of logs however, you need to visit the userlogs folder of every data node in the cluster, and aggregate the logs based on timestamp. This is because the name node dynamically picks which data nodes to execute a specific task during a job's execution. Figure 11-5 shows the userlogs directory of one of the data nodes after a few job executions in the cluster.

Figure 11-5. The userlogs folder

There are a few other log files that use the log4j framework. These log other cluster operations, specifically job executions. They are classified based on their respective project. For example:

hadoop.log: This file records only the MapReduce job execution output. Since it's the data nodes that actually carry out individual Map and Reduce tasks, this file is normally populated in the data nodes. It is found in the C:\apps\dist\hadoop-1.2.0.1.3.1.0-06\ logs directory.

templeton.log: This file logs the execution statistics of the jobs that are submitted using the Hadoop streaming interface. Job submissions using .NET SDK and PowerShell fall into this category. The log is available in the C:\apps\dist\hcatalog-0.11.0.1.3.1.0-06\logs folder.

hive.log: Found in the C:\apps\dist\hive-0.11.0.1.3.1.0-06\logs folder, this file records the output of all Hive job submissions. It is useful when a Hive job submission fails before even reaching the MapReduce phase.

oozie.log: Oozie web services streaming operations are logged to this file. It is present in the C:\apps\dist\oozie-3.3.2.1.3.1.0-06\oozie-win-distro\logs directory.

ooziejpa.log: Reports oozie database persistence level log messages. It is present in the C:\apps\dist\oozie-3.3.2.1.3.1.0-06\oozie-win-distro\logs directory.

oozieops.log: This file records all administrative tasks and operations messages for Oozie. It is present in the C:\apps\dist\oozie-3.3.2.1.3.1.0-06\oozie-win-distro\logs directory.

ooizeinstrumentation.log: This file records Oozie instrumentation data and is refreshed every 60 seconds. It is present in the C:\apps\dist\oozie-3.3.2.1.3.1.0-06\oozie-win-distro\logs directory.

pig_<Random_Number>.log: This file logs the results of Pig job executions from the Grunt shell. It is found in the C:\apps\dist\hadoop-1.2.0.1.3.1.0-06\logs folder.

Collectively, all these different types of log files will help you figure out issues in the event of a failure during service startup, job submission, or job execution.

Windows ODBC Tracing

One of the most common ways to consume HDInsight data is through Hive and the ODBC layer it exposes. The Windows operating system has built-in capabilities to trace all the ODBC driver API calls and their return values. Often, when client applications like Excel, Integration Services, and others fail to connect to HDInsight using the ODBC driver, the driver logging mechanism comes in handy.

Third-party ODBC drivers might not have built-in logging capability, for example the Microsoft Hive ODBC driver that is developed through partnership with Simba. In such scenarios you can use the standard ODBC logging mechanism from ODBC Data Source Administrator. The only difference here is that the standard mechanism is system-wide ODBC tracing for all ODBC drivers that are installed on your system as opposed to only the Hive ODBC driver.

▓ **Note** Enabling system-wide tracing from ODBC Data Source Administrator can significantly reduce performance of applications relying on ODBC function calls.

To enable system-wide ODBC tracing, launch the *ODBC Data Source Administrator* from *Control Panel* or click on Start ➤ Run ➤ odbcad32.exe. Navigate to the *Tracing* tab, and click on *Start Tracing Now* as shown in Figure 11-6.

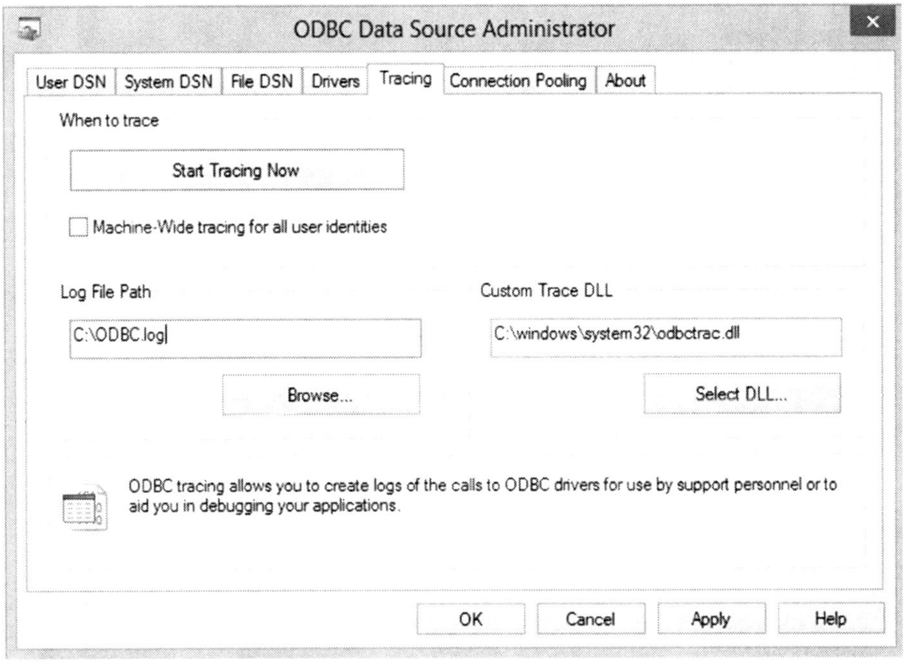

Figure 11-6. *Windows ODBC tracing*

You need to select the *Log File Path* to write the logs to. The *Custom Trace DLL* field should be pre-populated with the Windows-defined tracing dll and need not be changed. By default, it is set to the file `C:\windows\system32\odbctrac.dll`.

Once tracing is started, all subsequent ODBC function calls will be recorded in the log file in your local machine. A sample ODBC log file entries look similar to the following snippet:

```
test        1c4-186c    ENTER SQLAllocEnv
        HENV *              0x500BC504

test        1c4-186c    EXIT  SQLAllocEnv  with return code 0 SQL_SUCCESS)
        HENV *              0x500BC504 ( 0x008B9788)

test        1c4-186c    ENTER SQLAllocEnv
        HENV *              0x500BC508

test        1c4-186c    EXIT  SQLAllocEnv  with return code 0 (SQL_SUCCESS)
        HENV *              0x500BC508 ( 0x008B9808)

test        1c4-186c    ENTER SQLSetEnvAttr
        SQLHENV             0x008B9808
        SQLINTEGER          201 <SQL_ATTR_CONNECTION_POOLING>
        SQLPOINTER          0 <SQL_CP_OFF>
        SQLINTEGER               -6
```

```
test        1c4-186c      EXIT  SQLSetEnvAttr  with return code 0 (SQL_SUCCESS)
        SQLHENV               0x008B9808
        SQLINTEGER            201 <SQL_ATTR_CONNECTION_POOLING>
        SQLPOINTER            0 <SQL_CP_OFF>
        SQLINTEGER                 -6


test        1c4-186c      ENTER SQLAllocConnect
        HENV                  0x008B9808
        HDBC *                0x004CAAB8


test        1c4-186c      EXIT  SQLAllocConnect with return code 0(SQL_SUCCESS)
        HENV                  0x008B9808
        HDBC *                0x004CAAB8 ( 0x008BA108)


test        1c4-186c      ENTER SQLGetInfoW
        HDBC                  0x008BA108
        UWORD                      10 <SQL_ODBC_VER>
        PTR                   0x004CAA84
        SWORD                      22
        SWORD *               0x00000000


test        1c4-186c      EXIT  SQLGetInfoW  with return code 0 (SQL_SUCCESS)
        HDBC                  0x008BA108
        UWORD                      10 <SQL_ODBC_VER>
        PTR                   0x004CAA84 [      -3] "03.80.0000\ 0"
        SWORD                      22
        SWORD *               0x00000000
```

As you can see, the function calls are logged as a pair of *Enter* and *Exit* blocks along with the return codes. You can verify details like the connection pooling information and ODBC driver version from the trace. In the case of an error, you will see an error block with a diagnosis (*DIAG*) code for further analysis as in the following snippet:

```
test        1c4-186c      EXIT  SQLDriverConnectW  with return code -1 (SQL_ERROR)
        HDBC                  0x008BA108
        HWND                  0x000F0598
        WCHAR *               0x0F63B45C [      -3] "******\ 0"
        SWORD                      -3
        WCHAR *               0x0F63B45C
        SWORD                      -3
        SWORD *               0x00000000
        UWORD                       0 <SQL_DRIVER_NOPROMPT>


        DIAG [08001] Unable to establish connection with hive server (-1)
```

If the ODBC driver you use does not implement its own logging mechanism, this standard Windows ODBC trace is the only option to check ODBC API calls and their return codes. You can also follow the step-by-step process in the article at http://support.microsoft.com/kb/274551.

■ **Note** Make sure you turn off system-wide ODBC tracing once your data collection is over; otherwise, it can significantly hurt performance of the entire system. Data collection carries with it overhead that you should tolerate only when actively troubleshooting a problem.

Logging Windows Azure Storage Blob Operations

You can configure your storage account to monitor and log operations that pass over the Windows Azure Storage Blob (WASB). These include operations that you initiate from the Windows Azure Management Portal, from your .NET or PowerShell clients, as well as file system operations from the Hadoop Command Line. Because these operations might incur additional cost in terms of using storage space, logging and monitoring are turned off by default for your storage account. Monitoring and logging can be enabled on all three types of storage: blobs, tables, and queues. You can specify one of three available monitoring levels:

- Off
- Minimal
- Verbose

Similarly, you can set the logging activities on your storage to one of three levels:

- Read Requests
- Write Requests
- Delete Requests

Navigate to your storage account in the Azure Management portal. Click on the *Configure* link, and choose the desired level of logging, as shown in Figure 11-7.

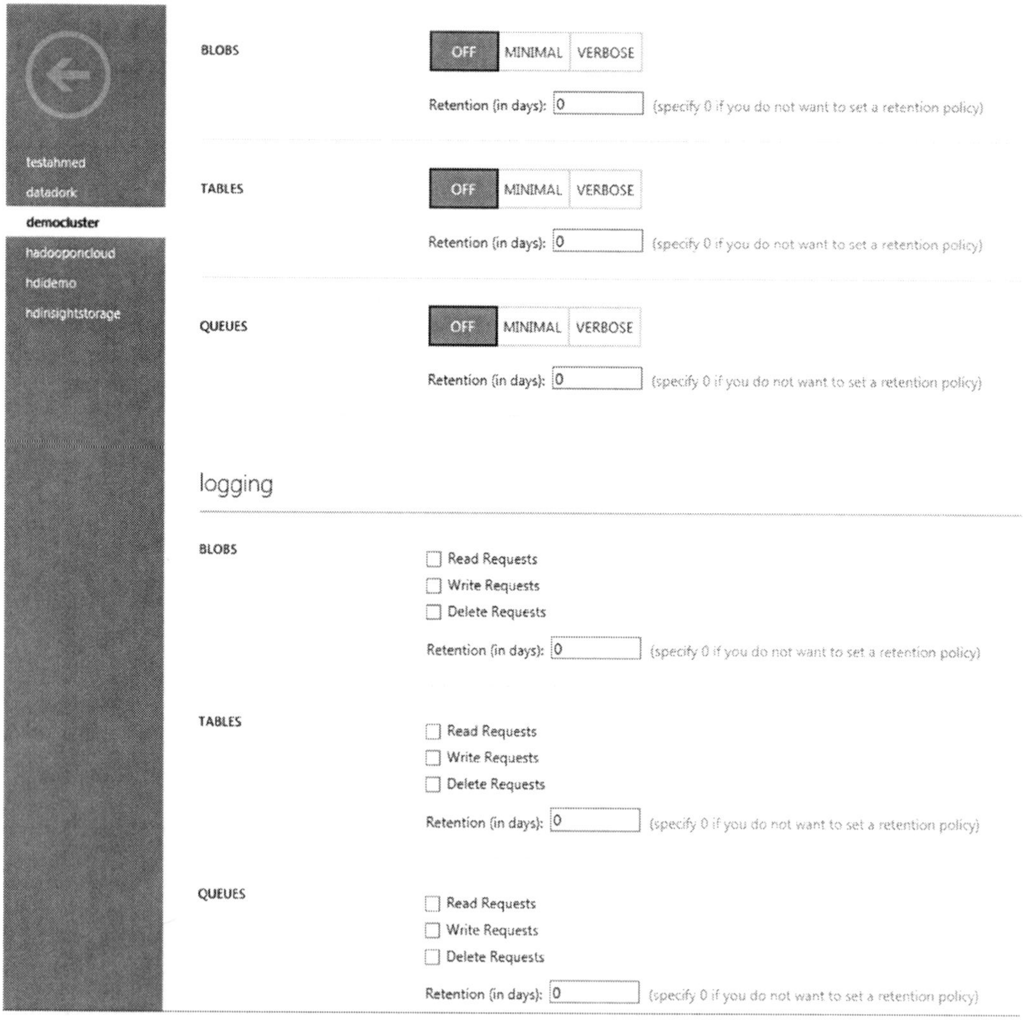

Figure 11-7. *Select monitoring and logging level*

Note that as you turn on verbose monitoring and logging, the Azure management portal warns you about the additional cost factor through visual clues and tool tips, as shown in Figure 11-8. Warning messages will have special icons, as well as a brightly-colored background to the text.

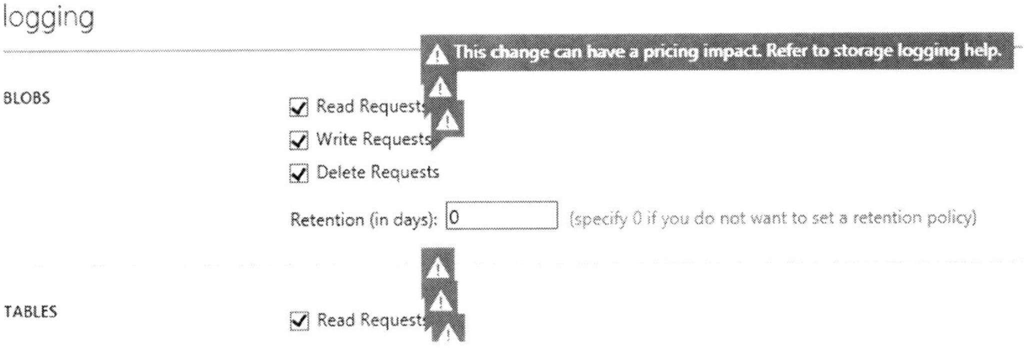

logging

Figure 11-8. *Pricing impact on logging and monitoring WASB*

Additionally, the Windows Azure's Logging infrastructure provides a trace of the executed requests against your storage account (blobs, tables, and queues). You can monitor requests made to your storage accounts, check the performance of individual requests, analyze the usage of specific containers and blobs, and debug storage APIs at a request level. To understand this logging infrastructure in depth and learn how to manage the storage analytics in detail, refer to the following blog post by the Azure Storage team:

http://blogs.msdn.com/b/windowsazurestorage/archive/2011/08/03/windows-azure-storage-logging-using-logs-to-track-storage-requests.aspx.

Logging in Windows Azure HDInsight Emulator

Windows Azure HDInsight Emulator is a single-node distribution of HDInsight available on Windows Server platforms. The logging mechanism on the emulator is almost exactly the same as in the Azure service. There are only some minor changes to the log file paths to worry about.

Basically, everything remains the same. The only real change is that the base directory changes to C:\Hadoop as opposed to the C:\apps\dist used in Azure. Also, since the emulator deploys HDInsight cluster version 1.6 as of this writing, the directory names of each of the projects also change. Figure 11-9 shows the directory structure of the emulator installation as of the writing of this book. There is every possibility that the emulator will match the Azure HDInsight cluster versions in the near future, and that everything will eventually be in sync.

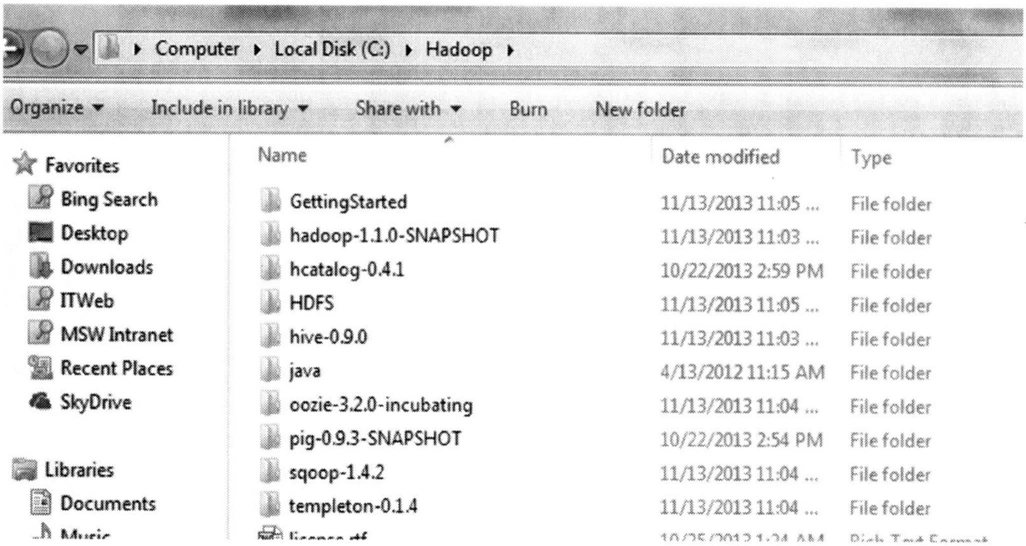

Figure 11-9. *The emulator directory structure*

■ **Note** The logging infrastructure changes in the emulator are explained in detail in Chapter 7.

Summary

This chapter walked you through the logging mechanism used in the HDInsight service. Although it focused on HDInsight-specific logging operations, it gives you a glimpse on how the traditional Apache Hadoop Log4j logging infrastructure can be leveraged as well. You read about several logging optimizations to avoid logging and maintaining irrelevant footprints. You also learned about enabling monitoring and logging on your Azure storage account through the Azure management portal.

Once an HDInsight cluster is operational and when it comes to consuming data, you need to know about logging the client-side driver calls as well. At the end of the day, data is viewed from interactive client applications like graphs and charting applications. Logging the Hive ODBC driver calls is very essential because it forms the bridge between your client consumer and your Hadoop cluster.

CHAPTER 12

■ ■ ■

Troubleshooting Cluster Deployments

Once you really start to play around with your HDInsight clusters, you are bound to end up with problems. Whether the problems are related to the manual or programmatic deployments of clusters or submitting your MapReduce jobs, troubleshooting is the art of logically removing the roadblocks that stand between you and your Big Data solution. This chapter will focus specifically on common cluster-deployment failure scenarios and ways to investigate them.

Cluster Creation

As you saw in Chapter 3, creating a cluster using either Quick Create or Custom Create involves a sequence of operations that need to be successfully completed to make the cluster operational. The phases are marked by the status shown at each stage:

- Submitting
- Accepted
- Windows Azure VM Configuration
- HDInsight Configuration
- Running

Table 12-1 explains what goes on behind the scenes during each of these phases.

Table 12-1. *Status designations when creating HDInsight clusters*

Status	What it means
Submitting	The communication in this step is between the Azure portal and the HDInsight Deployment Service, which is a REST API provided by HDInsight in Azure for its internal use in these kind of operations. If there is a failure here, it is likely a problem with the parameters of the setup, or a serious failure of the Deployment service.
Accepted	The HDInsight Deployment Service orchestrates the actions from this point forward, communicating status back to the Azure portal. A hidden Cloud Service is provisioned as a container, and then Cluster Storage is set up using your Storage Account and Key. A container is then created with a default name matching the storage account name. (Note: You can customize Storage Account links if you wish.) When this is successful, all the preconditions for setup have been met. If you encounter a failure at this step, it is highly likely that you have provided incorrect storage account details or duplicate container name.

(contuinued)

Table 12-1. (*continued*)

Status	What it means
Windows Azure VM Configuration	The HDInsight Deployment Service makes calls to Azure to initiate the provisioning of virtual machines (VMs) for the Head Node, Worker Nodes, and Gateway Node(s). The gateway acts as the security boundary between the cluster and the outside world. All traffic coming into the cluster goes through the gateway for authentication and authorization. The gateway can be thought of as a proxy that performs the necessary operations and forwards the request to the appropriate cluster components. So if you try to connect through the templeton or hive from, say, Excel, the call enters the gateway and then is proxied through to the rest of the components.
HDInsight Configuration	On startup, each node runs custom actions that download and install the appropriate components. These actions are coordinated by the individual node's local Deployment Agent. Installations of the Java Runtime, Hortonworks Data Platform, Microsoft HDInsight on Azure, and component bundles like Hive, Pig, Hcatalog, Sqoop and Oozie are run.
Running	The cluster is ready for use.

A few scenarios apart from the ones shown in the preceding table can lead to failure during the cluster-provisioning process:

- **Race condition exists on cluster creation.** An operation to create the hidden Cloud Service object was not synchronous. A subsequent call to retrieve the Cloud Service to use in the next step failed.

- **VM cores are limited by subscription.** Attempts to create a cluster using cores past the subscription limit failed.

- **Datacenter capacity is limited.** Because HDInsight clusters can use a large number of cores. Cluster creation failures can occur when the datacenter is near capacity.

- **Certain operations have must-succeed logging attached to them.** If the underlying logging infrastructure (Windows Azure Tables) is not available or times out, the cluster creation effort may fail.

Installer Logs

The Windows Azure HDInsight Service has a mechanism to log its cluster-deployment operations. Log files are placed in the *C:\HDInsightLogs* directory in the name node and data nodes. They contain two types of log files:

- AzureInstallHelper.log

- DeploymentAgent.log

These files give you information about several key aspects of the deployment process. Basically, after the VMs are provisioned, a deployment service runs for HDInsight that unpacks and installs Hadoop and its supporting projects with the necessary Windows services on the name node and data nodes. For example, if a node re-imaging has taken place, there will be re-imaging status entries at the very beginning of the *DeploymentAgent.log* file, as shown in Listing 12-1.

Listing 12-1. Node re-imaging status entries

```
11/10/2013 12:45:20 PM +00:00,11/10/2013 12:44:39 PM +00:00,
xxxxxxxxxxxxxxxxxxxxxxxxxxx,IsotopeWorkerNode,IsotopeWorkerNode_IN_1,4,
xxxxxxxxxxxxxxxxxx,2224,1020,SetupLogEntryEvent,1002,Info,null,MdsLogger,ClusterSetup,
Microsoft.Hadoop.Deployment.Engine,Azure reimaging state: 3 - REIMAGE_DATA_LOSS:
Services do not exist; Java directory does not exist.
Fresh installation.,1.0.0.0,xxxxxxxxxxxxxxxx,xxxxxxxxxxxxxx,False,null,null,null,
null,null,2013/11/10 12:44:39.480,Diagnostics,0000000000000000000,
0000000055834815115, 11/10/2013 12:44:00 PM +00:00
```

If there are any errors while deploying the Apache components or the services, due to some race condition while accessing the file system, you may see log entries in this file similar to Listing 12-2.

Listing 12-2. Error during deployment

```
Diagnostics Information: 1001 : OrigTS : 2013/07/20 21:47:21.109; EventId : 1002;
TraceLevel : Info; DeploymentEnvironment : ; ClassName : MdsLogger;
Component : ClusterSetup; ComponentAction : Microsoft.Hadoop.Deployment.Engine;
Details : Manually taking the backup off all files as the directory rename failed with exception
System.IO.IOException: The process cannot access the file because it is being used by another process.
    at System.IO.__Error.WinIOError(Int32 errorCode, String maybeFullPath)
    at System.IO.Directory.Move(String sourceDirName, String destDirName)
    at Microsoft.Hadoop.Deployment.Engine.Commands.AzureBeforeHadoopInstallCommand
      .Execute(DeploymentContext deploymentContext, INodeStore nodeStore); Version : 1.0.0.0;
      ActivityId : 8f270dd7-4691-4a69-945f-e0a1a81605c1; AzureVMName : RD00155D6135E3;
      IsException : False; ExceptionType : ; ExceptionMessage : ; InnerExceptionType : ;
      InnerExceptionMessage : ; Exception : ;
```

You may also come across scenarios where the cluster-creation process completes but you don't see the packages that should have been deployed to the nodes in place. For example, the cluster deployment is done but you don't find Hive installed in the *C:\Apps\Dist* directory. These installer and deployment logs could give you some insight if something went wrong after VM provisioning. In most of these cases, re-creating the cluster is the easiest and recommended solution.

For the HDInsight emulator, the same pair of deployment logs is generated, but in a different directory. They can be found in the *C:\HadoopInstallFiles* directory as shown in Figure 12-1.

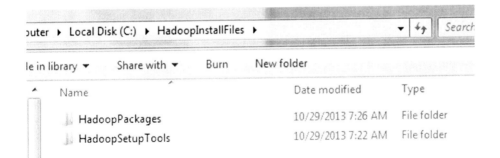

Figure 12-1. *HadoopInstallFiles directory*

The *HadoopPackages* folder contains the zipped Hortonworks Data Platform (HDP), which is basically a bundle of Hadoop and its supporting projects. The *HadoopSetupTools* folder contains the install/uninstall logs and the command files to initiate the installation or uninstallation. It also contains the command file and PowerShell script for invoking the packaged HDP from the *HadoopPackages* directory as shown in Figure 12-2.

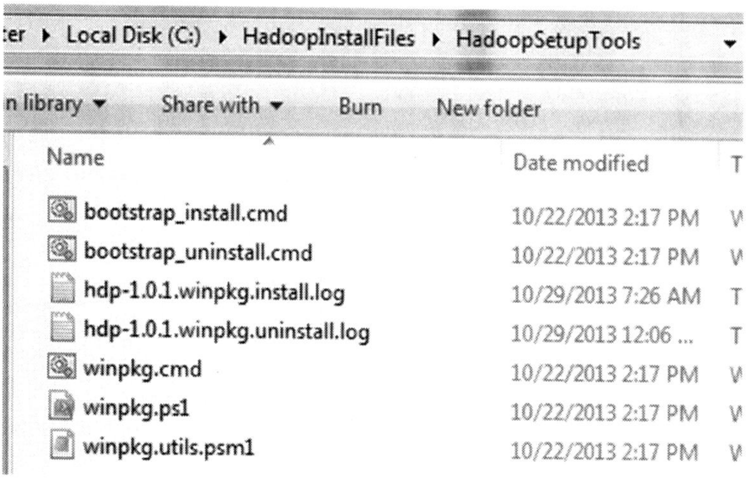

Figure 12-2. *HadoopSetupTools directory*

A typical `install.log` file contains the messages during the installation process sequentially as each project is deployed. A snippet of it looks similar to Listing 12-3.

Listing 12-3. HDInsight install log

```
WINPKG: Logging to existing log C:\HadoopInstallFiles\HadoopSetupTools\hdp-1.0.1.winpkg.install.log
WINPKG: ENV:WINPKG_BIN is C:\HadoopInstallFiles\HadoopSetupTools\
WINPKG: Setting Environment::CurrentDirectory to C:\Windows\SysWOW64
WINPKG: Current Directory: C:\Windows\SysWOW64
WINPKG: Package: C:\HadoopInstallFiles\HadoopSetupTools\\..\HadoopPackages\hdp-1.0.1-winpkg.zip
WINPKG: Action: install
WINPKG: Action arguments:
WINPKG: Run-WinpkgAction C:\HadoopInstallFiles\HadoopSetupTools\\..\HadoopPackages\
hdp-1.0.1-winpkg.zip C:\HadoopInstallFiles\HadoopPackages install
WINPKG: UNZIP: source C:\HadoopInstallFiles\HadoopPackages\hdp-1.0.1-winpkg.zip
WINPKG: UNZIP: destination C:\HadoopInstallFiles\HadoopPackages
WINPKG: UNZIP: unzipRoot C:\HadoopInstallFiles\HadoopPackages\hdp-1.0.1-winpkg
WINPKG: Unzip of C:\HadoopInstallFiles\HadoopPackages\
hdp-1.0.1-winpkg.zip to C:\HadoopInstallFiles\HadoopPackages succeeded
WINPKG: UnzipRoot: C:\HadoopInstallFiles\HadoopPackages\hdp-1.0.1-winpkg
WINPKG: C:\HadoopInstallFiles\HadoopPackages\hdp-1.0.1-winpkg\scripts\install.ps1
HDP: Logging to existing log C:\HadoopInstallFiles\HadoopSetupTools\hdp-1.0.1.winpkg.install.log
HDP: Logging to C:\HadoopInstallFiles\HadoopSetupTools\hdp-1.0.1.winpkg.install.log
HDP: HDP_INSTALL_PATH: C:\HadoopInstallFiles\HadoopPackages\hdp-1.0.1-winpkg\scripts
HDP: HDP_RESOURCES_DIR: C:\HadoopInstallFiles\HadoopPackages\hdp-1.0.1-winpkg\resources
HDP: INSTALLATION STARTED
HDP: Installing HDP @version@ to c:\hadoop
```

```
HDP: Installing Java
HDP: Installing Java
HDP: Logging to existing log C:\HadoopInstallFiles\HadoopSetupTools\hdp-1.0.1.winpkg.install.log
HDP: Logging to C:\HadoopInstallFiles\HadoopSetupTools\hdp-1.0.1.winpkg.install.log
HDP: HDP_INSTALL_PATH: C:\HadoopInstallFiles\HadoopPackages\hdp-1.0.1-winpkg\scripts
HDP: HDP_RESOURCES_DIR: C:\HadoopInstallFiles\HadoopPackages\hdp-1.0.1-winpkg\resources
HDP: Extracting Java archive into c:\hadoop
HDP: C:\HadoopInstallFiles\HadoopPackages\hdp-1.0.1-winpkg\resources\
winpkg.ps1 "C:\HadoopInstallFiles\HadoopPackages\hdp-1.0.1-winpkg\resources\java.zip"
utils unzip "c:\hadoop"
WINPKG: Logging to existing log C:\HadoopInstallFiles\HadoopSetupTools\
hdp-1.0.1.winpkg.install.log
WINPKG: ENV:WINPKG_BIN is C:\HadoopInstallFiles\HadoopPackages\
hdp-1.0.1-winpkg\resources
WINPKG: Setting Environment::CurrentDirectory to C:\HadoopInstallFiles\HadoopPackages\
hdp-1.0.1-winpkg\scripts
WINPKG: Current Directory: C:\HadoopInstallFiles\HadoopPackages\hdp-1.0.1-winpkg\scripts
WINPKG: Package: C:\HadoopInstallFiles\HadoopPackages\hdp-1.0.1-winpkg\resources\java.zip
WINPKG: Action: utils
WINPKG: Action arguments: unzip c:\hadoop
WINPKG: Run-BuiltInAction C:\HadoopInstallFiles\HadoopPackages\hdp-1.0.1-winpkg\
resources\java.zip C:\HadoopInstallFiles\HadoopPackages\hdp-1.0.1-winpkg\resources
utils unzip c:\hadoop
WINPKG: Preparing to unzip C:\HadoopInstallFiles\HadoopPackages\hdp-1.0.1-winpkg\
resources\java.zip to c:\hadoop
WINPKG: Finished processing C:\HadoopInstallFiles\HadoopPackages\hdp-1.0.1-winpkg\
resources\java.zip
HDP: Setting JAVA_HOME to c:\hadoop\java at machine scope
HDP: Done Installing Java
HDP: C:\HadoopInstallFiles\HadoopPackages\hdp-1.0.1-winpkg\scripts\
create_hadoop_user.ps1 -credentialFilePath c:\hadoop\singlenodecreds.xml
CREATE-USER: Logging to existing log C:\HadoopInstallFiles\HadoopSetupTools\
hdp-1.0.1.winpkg.install.log
CREATE-USER: Logging to C:\HadoopInstallFiles\HadoopSetupTools\hdp-1.0.1.winpkg.install.log
CREATE-USER: HDP_INSTALL_PATH: C:\HadoopInstallFiles\HadoopPackages\
hdp-1.0.1-winpkg\scripts
CREATE-USER: HDP_RESOURCES_DIR: C:\HadoopInstallFiles\HadoopPackages\
hdp-1.0.1-winpkg\resources
CREATE-USER: Username not provided. Using default username hadoop.
CREATE-USER: UserGroup not provided. Using default UserGroup HadoopUsers.
CREATE-USER: Password not provided. Generating a password.
CREATE-USER: Saving credentials to c:\hadoop\singlenodecreds.xml while running as FAREAST\desarkar
CREATE-USER: Creating user hadoop
CREATE-USER: User hadoop created
CREATE-USER: Granting SeCreateSymbolicLinkPrivilege
CREATE-USER: C:\HadoopInstallFiles\HadoopPackages\hdp-1.0.1-winpkg\resources\
installHelper2.exe -u PUMBAA\hadoop +r SeCreateSymbolicLinkPrivilege
CREATE-USER: SeCreateSymbolicLinkPrivilege granted
CREATE-USER: Granting SeServiceLogonRight
CREATE-USER: C:\HadoopInstallFiles\HadoopPackages\hdp-1.0.1-winpkg\resources\
installHelper2.exe -u PUMBAA\hadoop +r SeServiceLogonRight
```

```
CREATE-USER: Create user completed
CREATE-USER: Adding user to the local group
CREATE-USER: Group HadoopUsers successfully created
CREATE-USER: User hadoop successfully added to HadoopUsers.
HDP: Installing Hadoop Core
HDP: Setting HDFS_DATA_DIR to c:\hadoop\HDFS at machine scope
HDP: Invoke-Winpkg: C:\HadoopInstallFiles\HadoopPackages\hdp-1.0.1-winpkg\
resources\winpkg.ps1 "C:\HadoopInstallFiles\HadoopPackages\hdp-1.0.1-winpkg\
resources\hadoop-1.1.0-SNAPSHOT.winpkg.zip" install –credentialFilePath
c:\hadoop\singlenodecreds.xml -Verbose
WINPKG: Logging to existing log C:\HadoopInstallFiles\HadoopSetupTools\
hdp-1.0.1.winpkg.install.log
WINPKG: ENV:WINPKG_BIN is C:\HadoopInstallFiles\HadoopPackages\
hdp-1.0.1-winpkg\resources
WINPKG: Setting Environment::CurrentDirectory to C:\HadoopInstallFiles\HadoopPackages\
hdp-1.0.1-winpkg\scripts
WINPKG: Current Directory: C:\HadoopInstallFiles\HadoopPackages\hdp-1.0.1-winpkg\scripts
WINPKG: Package: C:\HadoopInstallFiles\HadoopPackages\hdp-1.0.1-winpkg\resources\
hadoop-1.1.0-SNAPSHOT.winpkg.zip
WINPKG: Action: install
WINPKG: Action arguments: -credentialFilePath c:\hadoop\singlenodecreds.xml
WINPKG: Run-WinpkgAction C:\HadoopInstallFiles\HadoopPackages\
hdp-1.0.1-winpkg\resources\hadoop-1.1.0-SNAPSHOT.winpkg.zip C:\HadoopInstallFiles\HadoopPackages\
hdp-1.0.1-winpkg\resources install -credentialFilePath c:\hadoop\singlenodecreds.xml
WINPKG: UNZIP: source C:\HadoopInstallFiles\HadoopPackages\hdp-1.0.1-winpkg\
resources\hadoop-1.1.0-SNAPSHOT.winpkg.zip
WINPKG: UNZIP: destination C:\HadoopInstallFiles\HadoopPackages\hdp-1.0.1-winpkg\resources
WINPKG: UNZIP: unzipRoot C:\HadoopInstallFiles\HadoopPackages\hdp-1.0.1-winpkg\
resources\hadoop-1.1.0-SNAPSHOT.winpkg
WINPKG: Unzip of C:\HadoopInstallFiles\HadoopPackages\hdp-1.0.1-winpkg\resources\
hadoop-1.1.0-SNAPSHOT.winpkg.zip to C:\HadoopInstallFiles\HadoopPackages\hdp-1.0.1-winpkg\resources
succeeded
WINPKG: UnzipRoot: C:\HadoopInstallFiles\HadoopPackages\hdp-1.0.1-winpkg\resources\
hadoop-1.1.0-SNAPSHOT.winpkg
WINPKG: C:\HadoopInstallFiles\HadoopPackages\hdp-1.0.1-winpkg\resources\
hadoop-1.1.0-SNAPSHOT.winpkg\scripts\install.ps1 -credentialFilePath c:\hadoop\singlenodecreds.xml
HADOOP: Logging to existing log C:\HadoopInstallFiles\HadoopSetupTools\hdp-1.0.1.winpkg.install.log
HADOOP: Logging to C:\HadoopInstallFiles\HadoopSetupTools\hdp-1.0.1.winpkg.install.log
HADOOP: HDP_INSTALL_PATH: C:\HadoopInstallFiles\HadoopPackages\
hdp-1.0.1-winpkg\resources\hadoop-1.1.0-SNAPSHOT.winpkg\scripts
HADOOP: HDP_RESOURCES_DIR: C:\HadoopInstallFiles\HadoopPackages\
hdp-1.0.1-winpkg\resources\hadoop-1.1.0-SNAPSHOT.winpkg\resources
HADOOP: nodeInstallRoot: c:\hadoop
HADOOP: hadoopInstallToBin: c:\hadoop\hadoop-1.1.0-SNAPSHOT\bin
HADOOP: Reading credentials from c:\hadoop\singlenodecreds.xml
HADOOP: Username: PUMBAA\hadoop
HADOOP: CredentialFilePath: c:\hadoop\singlenodecreds.xml
HADOOP: Stopping MapRed services if already running before proceeding with install
HADOOP: Stopping "mapreduce" "jobtracker tasktracker historyserver" services
HADOOP: Stopping jobtracker service
HADOOP: Stopping tasktracker service
```

```
HADOOP: Stopping historyserver service
HADOOP: Stopping HDFS services if already running before proceeding with install
HADOOP: Stopping "hdfs" "namenode datanode secondarynamenode" services
HADOOP: Stopping namenode service
HADOOP: Stopping datanode service
HADOOP: Stopping secondarynamenode service
HADOOP: Logging to existing log C:\HadoopInstallFiles\HadoopSetupTools\hdp-1.0.1.winpkg.install.log
..............
```

The installer log is a great place to review how each of the operations are set up and executed. Even if there are no errors during deployment, you should refer to this log for a detailed understanding of your own on the sequence of operations during the installation. The log is stripped off for brevity. It contains the messages for each of the projects that get deployed. I have stopped here at Hadoop; in your installer log, you would see the verbose message for Hive, Pig, Sqoop and the rest of the projects.

If there is a component missing after the installation (such as Hive), you can investigate the install.log file, scroll down to the section for the respective component, and track down the cause of the error.

Troubleshooting Visual Studio Deployments

As described in Chapter 4, you can use the Hadoop .NET SDK classes to programmatically deploy your HDInsight clusters through Microsoft Visual Studio projects. The Visual Studio IDE gives you a couple of great ways to debug your application when some operation throws errors or does not produce the desired output.

Using Breakpoints

A breakpoint is a special marker in your code that is active when executing the program while using the Visual Studio debugger. When the marker is reached, it causes the program to pause, changing the execution mode to break mode. You can then step through the code line by line using the Visual Studio debugging tools, while monitoring the contents of local and watched variables. You can set a breakpoint on a particular line from the *Debug* menu of Visual Studio or by simply pressing the function key *F9*. Figure 12-3 shows a sample scenario in your HadoopClient solution where a breakpoint is hit and you can examine your variable values.

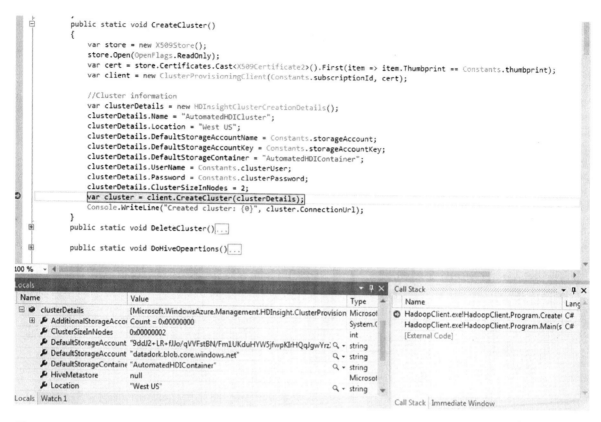

Figure 12-3. *Using breakpoints in Visual Studio*

Breakpoints are one of the most convenient ways to debug a program from Visual Studio. To learn more about setting, removing, and manipulating breakpoints, see the following MSDN article:

http://msdn.microsoft.com/en-us/library/5557y8b4(v=vs.90).aspx

---

■ **Note** Breakpoints are active only when using the Visual Studio debugger. When executing a program that has been compiled in release mode, or when the debugger is not active, breakpoints are unavailable.

Using IntelliTrace

IntelliTrace is a feature introduced in Visual Studio 2010 Ultimate that makes the life of a developer much easier when it comes to debugging. Visual Studio collects data about an application while it's executing to help developers diagnose errors. The collected data is referred to as *IntelliTrace events*.

These events are collected as part of the default debugging experience, and among other things, they let developers step back in time to see what happened in an application without having to restart the debugger. It is particularly useful when a developer needs a deeper understanding of code execution by providing a way to collect the complete execution history of an application.

Enable IntelliTrace for your application from the *Debug ➤ IntelliTrace ➤ Open IntelliTrace Settings* menu as shown in Figure 12-4.

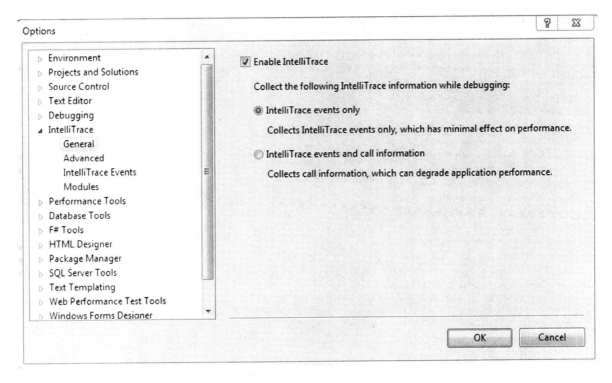

Figure 12-4. *Enabling IntelliTrace*

While you're debugging, IntelliTrace collects data about a managed application in the background, including information from many framework components such as ADO.NET, ASP.NET, and Hadoop.NET classes. When you break into the debugger, you are immediately presented with a sequential list of the IntelliTrace events that were collected. In your HadoopClient solution, if there is an error for which the cluster creation fails, you should see the errors in the sequence of events in the IntelliTrace events window as shown in Figure 12-5.

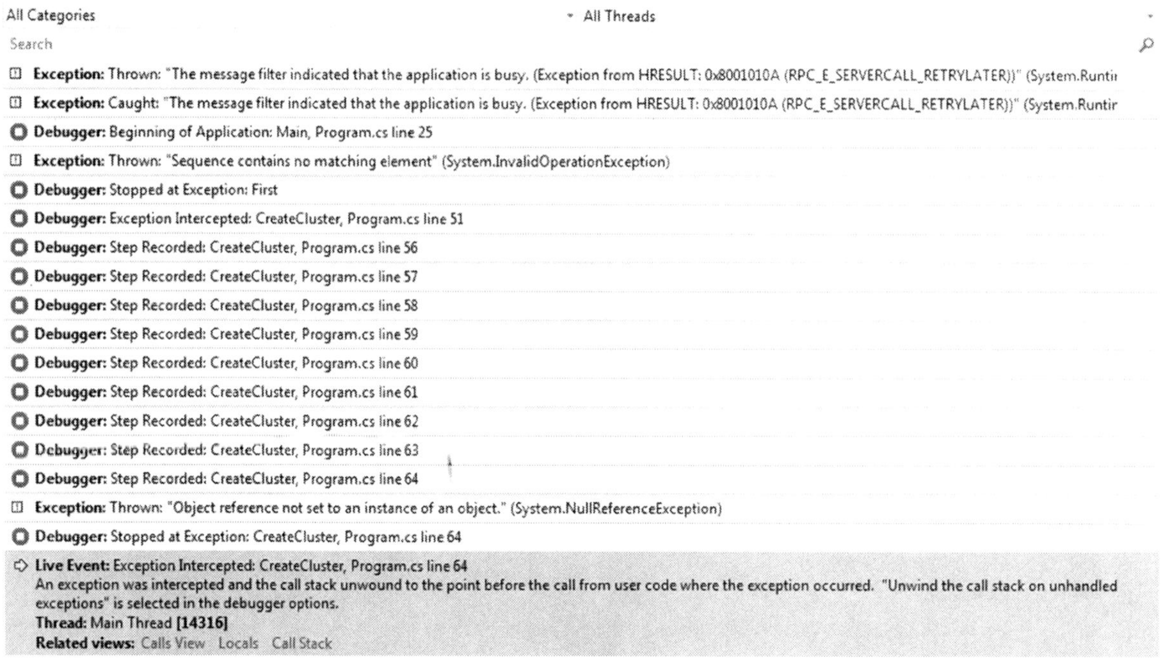

All Categories   ▾ All Threads   ▾

Search   ρ

▣ **Exception:** Thrown: "The message filter indicated that the application is busy. (Exception from HRESULT: 0x8001010A (RPC_E_SERVERCALL_RETRYLATER))" (System.Runti

▣ **Exception:** Caught: "The message filter indicated that the application is busy. (Exception from HRESULT: 0x8001010A (RPC_E_SERVERCALL_RETRYLATER))" (System.Runti

◉ **Debugger:** Beginning of Application: Main, Program.cs line 25

▣ **Exception:** Thrown: "Sequence contains no matching element" (System.InvalidOperationException)

◉ **Debugger:** Stopped at Exception: First

◉ **Debugger:** Exception Intercepted: CreateCluster, Program.cs line 51

◉ **Debugger:** Step Recorded: CreateCluster, Program.cs line 56

◉ **Debugger:** Step Recorded: CreateCluster, Program.cs line 57

◉ **Debugger:** Step Recorded: CreateCluster, Program.cs line 58

◉ **Debugger:** Step Recorded: CreateCluster, Program.cs line 59

◉ **Debugger:** Step Recorded: CreateCluster, Program.cs line 60

◉ **Debugger:** Step Recorded: CreateCluster, Program.cs line 61

◉ **Debugger:** Step Recorded: CreateCluster, Program.cs line 62

◉ **Debugger:** Step Recorded: CreateCluster, Program.cs line 63

◉ **Debugger:** Step Recorded: CreateCluster, Program.cs line 64

▣ **Exception:** Thrown: "Object reference not set to an instance of an object." (System.NullReferenceException)

◉ **Debugger:** Stopped at Exception: CreateCluster, Program.cs line 64

▷ **Live Event:** Exception Intercepted: CreateCluster, Program.cs line 64
An exception was intercepted and the call stack unwound to the point before the call from user code where the exception occurred. "Unwind the call stack on unhandled exceptions" is selected in the debugger options.
Thread: Main Thread **[14316]**
Related views: Calls View Locals Call Stack

Figure 12-5. *IntelliTrace events window*

If you opt to trace function-call sequences while enabling IntelliTrace, you can switch to the *IntelliTrace Calls View* and see the function calls as shown in Figure 12-6.

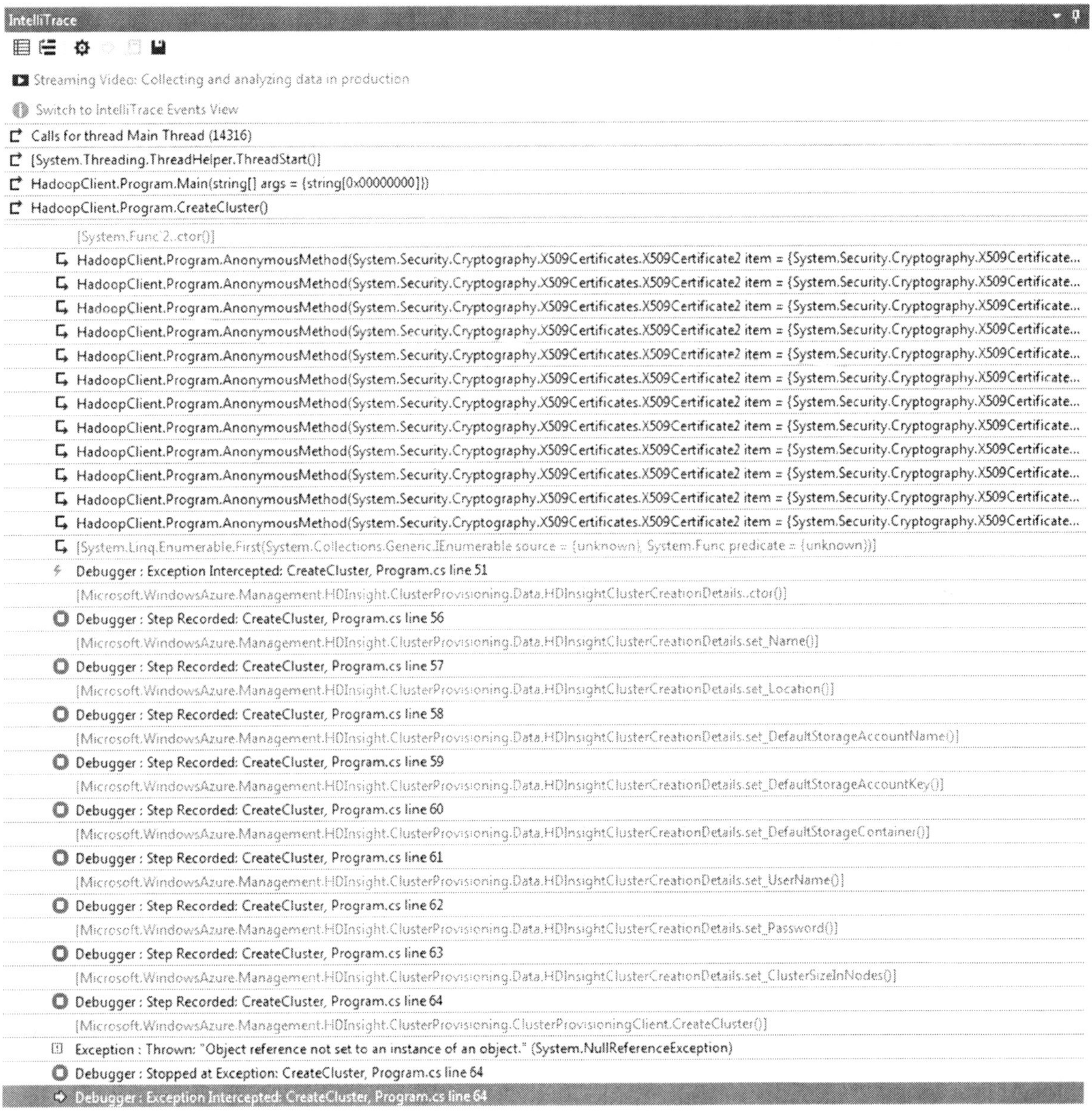

Figure 12-6. IntelliTrace calls view

Note that once you are in the calls view, the link in the IntelliTrace window toggles to *IntelliTrace Events View*.
IntelliTrace can greatly improve both your day-to-day development activities and your ability to quickly and easlly diagnose problems without having to restart your application and debug with the traditional break-step-inspect technique. This is just a brief overview of the feature. If you are interested, you can get more information about IntelliTrace at the following MSDN link.

```
http://msdn.microsoft.com/en-us/library/vstudio/dd286579.aspx
```

Troubleshooting PowerShell Deployments

Windows Azure PowerShell cmdlets provide another way to automate HDInsight cluster provisioning. Basically, you can use Windows PowerShell to perform a variety of tasks in Windows Azure, either interactively at a command prompt or automatically through scripts. Windows Azure PowerShell is a module that provides cmdlets to manage Windows Azure through Windows PowerShell.

You can use the cmdlets to create, test, deploy, and manage your HDInsight clusters on the Windows Azure platform. The following sections describe ways to enable the logging and debugging of PowerShell script executions, which can help you track down a cluster-deployment failure.

Using the *Write-** cmdlets

PowerShell has built-in cmdlets for logging that use the verb *write*. Each of the cmdlets is controlled by a shell variable that ends with *Preference* argument. For example, to turn the warning messages on, set the variable $*WarningPreference* to *Continue*.

Table 12-2 summarizes the different types of write cmdlets that PowerShell offers with the usage description for each.

Table 12-2. *PowerShell write-* cmdlets*

Cmdlet	Function
Write-Debug	The *Write-Debug* cmdlet writes debug messages to the console from a script or command.
Write-Error	The *Write-Error* cmdlet declares a nonterminating error. By default, errors are sent in the error stream to the host program to be displayed, along with output.
Write-EventLog	The *Write-EventLog* cmdlet writes an event to an event log. To write an event to an event log, the event log must exist on the computer and the source must be registered for the event log.
Write-Host	The *Write-Host* cmdlet customizes output. You can specify the color of text by using the ForegroundColor parameter, and you can specify the background color by using the BackgroundColor parameter. The Separator parameter lets you specify a string to use to separate displayed objects. The particular result depends on the program that is hosting Windows PowerShell.
Write-Output	The *Write-Output* cmdlet sends the specified object down the pipeline to the next command. If the command is the last command in the pipeline, the object is displayed in the console.
Write-Progress	The *Write-Progress* cmdlet displays a progress bar in a Windows PowerShell command window that depicts the status of a running command or script. You can select the indicators that the bar reflects and the text that appears above and below the progress bar.
Write-Verbose	The *Write-Verbose* cmdlet writes text to the verbose message stream in Windows PowerShell. Typically, the verbose message stream is used to deliver information about command processing that is used for debugging a command.
Write-Warning	The *Write-Warning* cmdlet writes a warning message to the Windows PowerShell host. The response to the warning depends on the value of the user's $WarningPreference variable and the use of the WarningAction common parameter.

Using the –debug Switch

Another option in PowerShell is to use the *–debug* switch while executing your scripts. This switch prints the status messages in the PowerShell command prompt during script execution and can help you debug your script failures. A sample output using the debug switch while trying to get cluster details with an incorrect subscription name is similar to the one shown in Listing 12-4.

Listing 12-4. The –debug switch

```
Get-AzureHDInsightCluster -Subscription incorrectsub -debug
DEBUG: Severity: Error
One or more errors occurred.
   at Microsoft.WindowsAzure.Management.HDInsight.Cmdlet.PSCmdlets.
   GetAzureHDInsightClusterCmdlet.EndProcessing()
One or more errors occurred.

Unable to resolve subscription 'incorrectsub'
   at Microsoft.WindowsAzure.Management.HDInsight.Cmdlet.
   GetAzureHDInsightClusters.AzureHDInsightCommandExtensions.
   ResolveSubscriptionId(String subscription)
   at Microsoft.WindowsAzure.Management.HDInsight.Cmdlet.
   GetAzureHDInsightClusters.AzureHDInsightCommandExtensions.
   GetSubscriptionCertificateCredentials(IAzureHDInsightCommonCommandBase command)
   at Microsoft.WindowsAzure.Management.HDInsight.Cmdlet.Commands.
   CommandImplementations.GetAzureHDInsightClusterCommand.<EndProcessing>d__2.MoveNext()
```

Summary

The Windows Azure HDInsight Service writes the sequence of installations during cluster deployments in specific log files. These log files are the ones to fall back to if your cluster-provisioning process encounters errors. Using a cloud service limits your control of the operations in compared to the control you have on your on-premises box products. This chapter taught you about troubleshooting mechanisms and places to start investigating when something goes wrong. You also learned about the different debugging mechanisms available with Visual Studio and Windows Azure PowerShell when provisioning your HDInsight clusters programmatically. In the next chapter, you will learn about troubleshooting the different types of job-submission failures in HDInsight.

CHAPTER 13

Troubleshooting Job Failures

There are different types of jobs you can submit to your HDInsight cluster, and it is inevitable that you will run into problems every now and then while doing so. Though most HDInsight jobs are internally executed as MapReduce jobs, there are different techniques for troubleshooting high-level supporting projects like Hive, Pig, Oozie, and others that make life easier for the developer. In this chapter, you will learn to troubleshoot the following types of failures:

- MapReduce job failures

- Hive job failures

- Pig job failures

- Sqoop job failures

- Windows Azure Storage Blob failures

- Cluster connectivity failures

MapReduce Jobs

All MapReduce job activities are logged by default in Hadoop in the C:\apps\dist\hadoop-1.2.0.1.3.1.0-06\logs\ directory of the name node. The log file name is of the format HADOOP-jobtracker-hostname.log. The most recent data is in the .log file; older logs have their date appended to them. In each of the Data Nodes or Task Nodes, you will also find a subdirectory named userlogs inside the C:\apps\dist\hadoop-1.2.0.1.3.1.0-06\logs\ folder.

This directory will have another subdirectory for every MapReduce task running in your Hadoop cluster. Each task records its stdout (output) and stderr (error) to two files in this subdirectory. If you are running a multinode Hadoop cluster, the logs you will find here are not centrally aggregated. To put together a complete picture, you will need to check and verify each Task Node's /logs/userlogs/ directory for their output, and then create the full log history to understand what went wrong in a particular job.

In a Hadoop cluster, the entire job submission, execution, and history-management process is done by three types of services:

- **JobTracker**   JobTracker is the master of the system, and it manages the jobs and resources in the cluster (TaskTrackers). The JobTracker schedules and coordinates with each of the TaskTrackers that are launched to complete the jobs.

- **TaskTrackers**   These are the slave services deployed on Data Nodes or Task Nodes. They are responsible for running the map and reduce tasks as instructed by the JobTracker.

- **JobHistoryServer**   This is a service that serves historical information about completed jobs. JobHistoryServer can be embedded within the JobTracker process. If you have an extremely busy cluster, it is recommended that you run this as a separate service. This can be done by setting the `mapreduce.history.server.embedded` property to `true` in the `mapred-site.xml` file. Running this service consumes considerable disk space because it saves job history information for all the jobs.

■ **Note**   In Hadoop versions 2.0 and beyond, MapReduce will be replaced by *YARN* or MapReduce 2.0 (also known as MRv2). YARN is a subproject of Hadoop at the Apache Software Foundation that was introduced in Hadoop 2.0. It separates the resource-management and processing components. It provides a more generalized processing platform that is not restricted to just MapReduce.

Configuration Files

There are two key configuration files that have the various parameters for MapReduce jobs. These files are located in the path `C:\apps\dist\hadoop-1.2.0.1.3.1.0-06\conf\` of the NameNode:

- core-site.xml

- mapred-site.xml

core-site.xml

This file contains configuration settings for Hadoop Core, such as I/O settings that are common to Windows Azure Storage Blob (WASB) and MapReduce. It is used by all Hadoop services and clients because all services need to know how to locate the NameNode. There will be a copy of this file in each node running a Hadoop service. This file has several key elements of interest—particularly because the storage infrastructure has moved to WASB instead of being in Hadoop Distributed File System (HDFS), which used to be local to the data nodes. For example, in your democluster, you should see entries in your `core-site.xml` file similar to Listing 13-1.

Listing 13-1. WASB detail

```
<property>
    <name>fs.default.name</name>
    <!-- cluster variant -->
    <value>wasb://democlustercontainer@democluster.blob.core.windows.net
        </value>
    <description>The name of the default file system.  Either the
        literal string "local" or a host:port for NDFS.
        </description>
    <final>true</final>
</property>
```

If there is an issue with accessing your storage that is causing your jobs to fail, the `core-site.xml` file is the first place where you should confirm that your cluster is pointing toward the correct storage account and container.

The `core-site.xml` file also has an attribute for the storage key, as shown in Listing 13-2. If you are encountering *502/403 - Forbidden/Authentication* errors while accessing your storage, you must make sure that the proper storage account key is provided.

Listing 13-2. Storage account key

```
<property>
    <name>fs.azure.account.key.democluster.blob.core.windows.net</name>
    <value>YourStorageAccountKey</value>
</property>
```

There are also several Azure throttling factors and blob IO buffer parameters that can be set through the core-site.xml file. They are outlined in Listing 13-3.

Listing 13-3. Azure throttling factors

```
<property>
    <name>fs.azure.selfthrottling.write.factor</name>
    <value>1.000000</value>
</property>

<property>
    <name>fs.azure.selfthrottling.read.factor</name>
    <value>1.000000</value>
</property>

<property>
    <name>fs.azure.buffer.dir</name>
    <value>/tmp</value>
</property>

<property>
    <name>io.file.buffer.size</name>
    <value>131072</value>
</property>
```

■ **Note**   Azure throttling is discussed in the section "Windows Azure Storage" later in this chapter.

mapred-site.xml

The mapred-site.xml file has the configuration settings for MapReduce services. It contains parameters for the JobTracker and TaskTracker processes. These parameters determine where the MapReduce jobs place their intermediate files and control files, the virtual memory usage by the Map and Reduce jobs, the maximum numbers of mappers and reducers, and many such settings.

In the case of a poorly performing job, optimizations such as moving the intermediate files to a fast Redundant Array of Inexpensive Disks (RAID) can be really helpful. Also, in certain scenarios when you know your job well, you may want to control the number of mappers or reducers being spawned for your job or increase the default timeout that is set for Map jobs. Listing 13-4 shows a few of the important attributes in mapred-site.xml.

Listing 13-4. mapred-site.xml

```
<property>
    <name>mapred.tasktracker.map.tasks.maximum</name>
    <value>4</value>
</property>

<property>
    <name>mapred.tasktracker.reduce.tasks.maximum</name>
    <value>2</value>
</property>

<property>
    <name>mapred.map.max.attempts</name>
    <value>8</value>
</property>

<property>
    <name>mapred.reduce.max.attempts</name>
    <value>8</value>
</property>

<property>
    <name>mapred.task.timeout</name>
    <value>600000</value>
  </property>

<property>
    <name>mapred.max.split.size</name>
    <value>536870912</value>
</property>
```

If you have active Hadoop clusters, there are numerous scenarios in which you have to come back and check the properties in Listing 13-4. Most of these properties come into the picture when there are job optimization or tuning requirements that cause jobs to take an unusually long time to complete. For several other types of obvious errors that may occur during a job submission, the log files can be a source of a great deal of information.

Log Files

I covered the different types of logs generated by Hadoop and the HDInsight service in detail in Chapter 11. However, let's go quickly through the logging infrastructure for MapReduce jobs again. The log files are normally stored in `C:\apps\dist\hadoop-1.2.0.1.3.1.0-06\logs\` and `C:\apps\dist\hadoop-1.2.0.1.3.1.0-06\bin\` folders by default. The `jobtracker.trace.log` file resides in the bin directory, and it logs the job startup command and the process id. A sample trace would be similar to Listing 13-5.

Listing 13-5. jobtracker.trace.log

```
HadoopServiceTraceSource Information: 0 : Tracing successfully initialized
    DateTime=2013-11-24T06:35:12.0190000Z
    Timestamp=3610300511
HadoopServiceTraceSource Information: 0 : Loading service xml:
    c:\apps\dist\hadoop-1.2.0.1.3.1.0-06\bin\jobtracker.xml
```

```
        DateTime=2013-11-24T06:35:12.0190000Z
        Timestamp=3610344009
HadoopServiceTraceSource Information: 0 :
        Successfully parsed service xml for service jobtracker
        DateTime=2013-11-24T06:35:12.0190000Z
        Timestamp=3610353933
HadoopServiceTraceSource Information: 0 : Command line:
        c:\apps\dist\java\bin\java -server -Xmx4096m  -Dhadoop.log.dir=
        c:\apps\dist\hadoop-1.2.0.1.3.1.0-06\logs -Dhadoop.log.file=
        hadoop-jobtracker-RD00155D67172B.log -Dhadoop.home.dir=
        c:\apps\dist\hadoop-1.2.0.1.3.1.0-06 -Dhadoop.root.logger=
        INFO,console,DRFA,ETW,FilterLog -Djava.library.path=;
        c:\apps\dist\hadoop-1.2.0.1.3.1.0-06\lib\nativc\Windows_NT-amd64-64;
        c:\apps\dist\hadoop-1.2.0.1.3.1.0-06\lib\native -Dhadoop.policy.file=
        hadoop-policy.xml -Dcom.sun.management.jmxremote  -Detwlogger.component=
        jobtracker  -Dwhitelist.filename=core-whitelist.res  -classpath
        c:\apps\dist\hadoop-1.2.0.1.3.1.0-6\conf;c:\apps\dist\java\lib\tools.jar;
        c:\apps\dist\hadoop-1.2.0.1.3.1.0-06;c:\apps\dist\hadoop-1.2.0.1.3.1.0-06\
        hadoop-ant-1.2.0.1.3.1.0-06.jar;c:\apps\dist\hadoop-1.2.0.1.3.1.0-06\
        hadoop-client-1.2.0.1.3.1.0-06.jar;c:\apps\dist\hadoop-1.2.0.1.3.1.0-06\
        hadoop-core-1.2.0.1.3.1.0-06.jar;c:\apps\dist\hadoop-1.2.0.1.3.1.0-06\
        hadoop-core.jar;c:\apps\dist\hadoop-1.2.0.1.3.1.0-06\
        hadoop-examples-1.2.0.1.3.1.0-06.jar;c:\apps\dist\
        hadoop-1.2.0.1.3.1.0-06\hadoop-examples.jar;c:\apps\dist\
        hadoop-1.2.0.1.3.1.0-06\hadoop-minicluster-1.2.0.1.3.1.0-06.jar;
        c:\apps\dist\hadoop-1.2.0.1.3.1.0-06\hadoop-test-1.2.0.1.3.1.0-06.jar;
        c:\apps\dist\hadoop-1.2.0.1.3.1.0-06\hadoop-test.jar;
        c:\apps\dist\hadoop-1.2.0.1.3.1.0-06\hadoop-tools-1.2.0.1.3.1.0-06.jar;
        c:\apps\dist\hadoop-1.2.0.1.3.1.0-06\hadoop-tools.jar;
        c:\apps\dist\hadoop-1.2.0.1.3.1.0-06\lib\*;
        c:\apps\dist\hadoop-1.2.0.1.3.1.0-06\lib\jsp-2.1\*;
        c:\apps\dist\log4jetwappender\microsoft-log4j-etwappender-1.0.jar;
org.apache.hadoop.mapred.JobTracker
        DateTime=2013-11-24T06:35:12.0190000Z
        Timestamp=3610354520
HadoopServiceTraceSource Information: 0 : ServiceHost#OnStart
        DateTime=2013-11-24T06:35:12.0346250Z
        Timestamp=3610410266
HadoopServiceTraceSource Information: 0 : Child process started, PID: 4976
        DateTime=2013-11-24T06:35:12.0346250Z
        Timestamp=3610428330
```

Apart from the trace file, Hadoop has built-in logging mechanisms implementing the log4j framework. The following JobTracker log files are located in the C:\apps\dist\hadoop-1.2.0.1.3.1.0-06\logs\ folder:

- hadoop-jobtracker<Hostname>.log

- hadoop-tasktracker<Hostname>.log

- hadoop-historyserver<Hostname>.log

These files record the actual execution status of the MapReduce jobs. Listing 13-6 shows an excerpt of the JobTracker log just after a MapReduce job is started.

Listing 13-6. Hadoop JobTracker Log

```
2013-11-24 06:35:12,972 INFO org.apache.hadoop.mapred.JobTracker:
STARTUP_MSG:
/*************************************************************
STARTUP_MSG: Starting JobTracker
STARTUP_MSG:    host = RD00155XXXXXX/xxx.xx.xx.xx
STARTUP_MSG:    args = []
STARTUP_MSG:    version = 1.2.0.1.3.1.0-06
STARTUP_MSG:    build = git@github.com:hortonworks/hadoop-monarch.git on branch
(no branch) -r f4cb3bb77cf3cc20c863de73bd6ef21cf069f66f; compiled by 'jenkins'
on Wed Oct 02 21:38:25 Coordinated Universal Time 2013
STARTUP_MSG:    java = 1.7.0-internal
*************************************************************/
2013-11-24 06:35:13,925 WARN org.apache.hadoop.metrics2.impl.MetricsSystemImpl:
Source name ugi already exists!
2013-11-24 06:35:13,925 INFO org.apache.hadoop.security.token.delegation.
AbstractDelegationTokenSecretManager:
Updating the current master key for generating delegation tokens
2013-11-24 06:35:13,940 INFO org.apache.hadoop.mapred.JobTracker:
Scheduler configured with (memSizeForMapSlotOnJT, memSizeForReduceSlotOnJT,
limitMaxMemForMapTasks, limitMaxMemForReduceTasks) (-1, -1, -1, -1)
2013-11-24 06:35:14,347 INFO org.apache.hadoop.http.HttpServer: listener.getLocalPort()
returned 50030 webServer.getConnectors()[0].getLocalPort() returned 50030
2013-11-24 06:35:14,362 INFO org.apache.hadoop.http.HttpServer:
Jetty bound to port 50030
2013-11-24 06:35:16,264 INFO org.apache.hadoop.mapred.JobTracker:
Setting safe mode to false. Requested by : hdp
2013-11-24 06:35:16,329 INFO org.apache.hadoop.util.NativeCodeLoader:
Loaded the native-hadoop library
2013-11-24 06:35:16,387 INFO org.apache.hadoop.mapred.JobTracker:
Cleaning up the system directory
2013-11-24 06:35:17,172 INFO org.apache.hadoop.mapred.JobHistory:
Creating DONE folder at
wasb://democlustercontainer@democluster.blob.core.windows.net/mapred/history/done
2013-11-24 06:35:17,536 INFO org.apache.hadoop.mapred.JobTracker:
History server being initialized in embedded mode
2013-11-24 06:35:17,555 INFO org.apache.hadoop.mapred.JobHistoryServer:
Started job history server at: 0.0.0.0:50030
Adding a new node: /fd0/ud0/workernode0
2013-11-24 06:35:18,363 INFO org.apache.hadoop.mapred.JobTracker:
Adding tracker tracker_workernode0:127.0.0.1/127.0.0.1:49186 to host workernode0
2013-11-24 06:35:19,083 INFO org.apache.hadoop.net.NetworkTopology:
Adding a new node: /fd1/ud1/workernode1
2013-11-24 06:35:19,094 INFO org.apache.hadoop.mapred.JobTracker:
Adding tracker tracker_workernode1:127.0.0.1/127.0.0.1:49193 to host workernode1
2013-11-24 06:35:19,365 INFO org.apache.hadoop.mapred.CapacityTaskScheduler:
Initializing 'joblauncher' queue with cap=25.0, maxCap=25.0, ulMin=100,
ulMinFactor=100.0, supportsPriorities=false, maxJobsToInit=750, maxJobsToAccept=7500,
 maxActiveTasks=200000, maxJobsPerUserToInit=750,
maxJobsPerUserToAccept=7500, maxActiveTasksPerUser=100000
2013-11-24 06:35:19,367 INFO org.apache.hadoop.mapred.CapacityTaskScheduler:
```

```
Initializing 'default' queue with cap=75.0, maxCap=-1.0, ulMin=100, ulMinFactor=100.0,
supportsPriorities=false, maxJobsToInit=2250, maxJobsToAccept=22500,
maxActiveTasks=200000, maxJobsPerUserToInit=2250,
maxJobsPerUserToAccept=22500, maxActiveTasksPerUser=100000
2013-11-24 07:05:16,099 INFO org.apache.hadoop.mapred.JobTracker:
jobToken generated and stored with users keys in /mapred/system/job_201311240635_0001/jobToken
2013-11-24 07:05:16,796 INFO org.apache.hadoop.mapred.JobInProgress:
job_201311240635_0001: nMaps=1 nReduces=0 max=-1
2013-11-24 07:05:16,799 INFO org.apache.hadoop.mapred.JobQueuesManager:
Job job_201311240635_0001 submitted to queue joblauncher
2013-11-24 07:05:16,800 INFO org.apache.hadoop.mapred.JobTracker:
Job job_201311240635_0001 added successfully for user 'admin' to queue 'joblauncher'
2013-11-24 07:05:16,803 INFO org.apache.hadoop.mapred.AuditLogger: USER=admin
IP=xx.xx.xx.xx    OPERATION=SUBMIT_JOB    TARGET=job_201311240635_0001    RESULT=SUCCESS
2013-11-24 07:05:19,329 INFO org.apache.hadoop.mapred.JobInitializationPoller:
Passing to Initializer Job Id :job_201311240635_0001 User: admin Queue : joblauncher
2013-11-24 07:05:24,324 INFO org.apache.hadoop.mapred.JobInitializationPoller:
Initializing job : job_201311240635_0001 in Queue joblauncher For user : admin
2013-11-24 07:05:24,324 INFO org.apache.hadoop.mapred.JobTracker:
Initializing job_201311240635_0001
2013-11-24 07:05:24,325 INFO org.apache.hadoop.mapred.JobInProgress:
Initializing job_201311240635_0001
2013-11-24 07:05:24,576 INFO org.apache.hadoop.mapred.JobInProgress:
Input size for job job_201311240635_0001 = 0. Number of splits = 1
2013-11-24 07:05:24,577 INFO org.apache.hadoop.mapred.JobInProgress:
job_201311240635_0001 LOCALITY_WAIT_FACTOR=0.0
2013-11-24 07:05:24,578 INFO org.apache.hadoop.mapred.JobInProgress:
Job job_201311240635_0001 initialized successfully with 1 map tasks and 0 reduce tasks.
2013-11-24 07:05:24,659 INFO org.apache.hadoop.mapred.JobTracker:
Adding task (JOB_SETUP) 'attempt_201311240635_0001_m_000002_0' to tip
task_201311240635_0001_m_000002, for tracker 'tracker_workernode1:127.0.0.1/127.0.0.1:49193'
2013-11-24 07:05:28,224 INFO org.apache.hadoop.mapred.JobInProgress:
Task 'attempt_201311240635_0001_m_000002_0' has completed task_201311240635_0001_m_000002 successfully.
```

The highlighted sections of the preceding log gives you the key settings configured to execute this job. Because the jobtracker.trace.log file records the command, you can easily figure out which of the parameters are overridden in the command line and which are the ones being inherited from the configuration files and then take appropriate corrective actions.

Compress Job Output

Hadoop is intended for storing large data volumes, so compression becomes a mandatory requirement. You can choose to compress your MapReduce job output by adding the following two parameters in your mapred-site.xml file:

```
mapred.output.compress=true
mapred.output.compression.codec= com.hadoop.compression.GzipCodec
```

Apart from these parameters, MapReduce provides facilities for the application developer to specify compression for both intermediate map outputs and the job outputs—that is, the output of the reducers. Such compression can be set up with CompressionCodec class implementation for the zlib compression algorithm in your custom MapReduce program. For extensive details on Hadoop compression, see the whitepaper http://msdn.microsoft.com/en-us/dn168917.aspx.

Concatenate Input Files

Concatenation is another technique that can improve your MapReduce job performance. The MapReduce program is designed to handle few larger files well in comparison to several smaller files. Thus, you can concatenate many small files into a few larger ones. This needs to be done in the program code where you implement your own MapReduce job. MapReduce can concatenate multiple small files to make it one block size, which is more efficient in terms of storage and data movement.

Avoid Spilling

All data in a Hadoop MapReduce job is handled as key-value pairs. All input data received by the user-defined method that constitutes the reduce task is guaranteed to be sorted by key. This sorting happens in two parts. The first sorting happens local to each mapper as the mapper reads the input data from one or more splits and produces the output from the mapping phase. The second sorting happens after a reducer has collected all the data from one or more mappers, and then produces the output from the shuffle phase.

The process of *spilling* during the map phase is the phenomenon in which complete input to the mapper cannot be held in memory before the final sorting can be performed on the output from the mapper. As each mapper reads input data from one or more splits, the mapper requires an in-memory buffer to hold the unsorted data as key-value pairs. If the Hadoop job configuration is not optimized for the type and size of the input data, the buffer can get filled up before the mapper has finished reading its data. In that case, the mapper will sort the data already in the filled buffer, partition that data, serialize it, and write (spill) it to the disk. The result is referred to as a *spill file*.

Separate spill files are created each time a mapper has to spill data. Once all the data has been read and spilled, the mapper will read all the spilled files again, sort and merge the data, and write (spill) that data back into a single file known as an *attempt file*.

If there is more than one spill, there must be one extra read and write of the entire data. So there will be three times (3x) the required I/O during the mapping phase, a phenomenon known as *data I/O explosion*. The goal is to spill only once (1x) during the mapping phase, which is a goal that can be achieved only if you carefully select the correct configuration for your Hadoop MapReduce job.

The memory buffer per-data record consists of three parts. The first part is the offset of the data record stored as a tuple. That tuple requires 12 bytes per record, and it contains the partition key, the key offset, and a value offset. The second part is the indirect sort index, requiring four bytes per record. Together, these two parts constitute the metadata for a record, for a total of 16 bytes per record. The third part is the record itself, which is the serialized key-value pair requiring R bytes, where R is the number of bytes of data.

If each mapper handles N records, the recommended value of the parameter that sets the proper configuration in the mapred-site.xml is expressed as follows:

```
<property>
    <name>io.sort.mb</name><value>N*(16+R)/(1024*1024)</value>
</property>
```

By specifying your configuration in this way, you reduce the chance of unwanted spill operations.

Hive Jobs

The best place to start looking at a Hive command failure is the Hive log file, which can be configured by editing the hive-site.xml file. The location of the hive-site.xml file is the C:\apps\dist\hive-0.11.0.1.3.0.1-0302\conf\ directory. Listing 13-7 is a sample snippet that shows how you can specify the Hive log file path.

Listing 13-7. hive-site.xml

```
<property>
    <name>hive.log.dir</name>
    <value>c:\apps\dist\hive-0.11.0.1.3.0.1-0302\logs</value>
</property>
```

Listing 13-7 shows the default location of the log file for Hive which is the \logs folder. The log file is created with the name of hive.log.

Log Files

Any *Data Definition Language* (*DDL*) or *Data Manipulation Language* (*DML*) commands are logged in the log files. For example, if you execute an HQL, CREATE DATABASE TEST and it gets created successfully, you should see similar entries in your hive.log file as shown in Listing 13-8.

Listing 13-8. hive.log

```
2013-11-15 11:56:49,326 INFO  ql.Driver (PerfLogger.java:PerfLogBegin(100))
- <PERFLOG method=Driver.run>
2013-11-15 11:56:49,326 INFO  ql.Driver (PerfLogger.java:PerfLogBegin(100))
- <PERFLOG method=TimeToSubmit>
2013-11-15 11:56:49,326 INFO  ql.Driver (PerfLogger.java:PerfLogBegin(100))
- <PERFLOG method=compile>
2013-11-15 11:56:49,327 INFO  parse.ParseDriver (ParseDriver.java:parse(179))
- Parsing command: create database test
2013-11-15 11:56:49,329 INFO  parse.ParseDriver (ParseDriver.java:parse(197))
- Parse Completed
2013-11-15 11:56:49,331 INFO  ql.Driver (Driver.java:compile(442))
- Semantic Analysis Completed
2013-11-15 11:56:49,332 INFO  ql.Driver (Driver.java:getSchema(259))
- Returning Hive schema: Schema(fieldSchemas:null, properties:null)
2013-11-15 11:56:49,332 INFO  ql.Driver (PerfLogger.java:PerfLogEnd(127))
- </PERFLOG method=compile start=1384516609326 end=1384516609332 duration=6>
2013-11-15 11:56:49,332 INFO  ql.Driver (PerfLogger.java:PerfLogBegin(100))
- <PERFLOG method=Driver.execute>
2013-11-15 11:56:49,333 INFO  ql.Driver (Driver.java:execute(1066))
- Starting command: create database test
2013-11-15 11:56:49,333 INFO  ql.Driver (PerfLogger.java:PerfLogEnd(127))
- </PERFLOG method=TimeToSubmit start=1384516609326 end=1384516609333 duration=7>
2013-11-15 11:56:49,871 INFO  ql.Driver (PerfLogger.java:PerfLogEnd(127))
- </PERFLOG method=Driver.execute start=1384516609332 end=1384516609871 duration=539>
2013-11-15 11:56:49,872 INFO  ql.Driver (SessionState.java:printInfo(423))
- OK
2013-11-15 11:56:49,872 INFO  ql.Driver (PerfLogger.java:PerfLogBegin(100))
- <PERFLOG method=releaseLocks>
2013-11-15 11:56:49,872 INFO  ql.Driver (PerfLogger.java:PerfLogEnd(127))
- </PERFLOG method=releaseLocks start=1384516609872 end=1384516609872 duration=0>
2013-11-15 11:56:49,872 INFO  ql.Driver (PerfLogger.java:PerfLogEnd(127))
- </PERFLOG method=Driver.run start=1384516609326 end=1384516609872 duration=546>
2013-11-15 11:56:49,873 INFO  CliDriver (SessionState.java:printInfo(423))
```

```
- Time taken: 0.548 seconds
2013-11-15 11:56:49,874 INFO  ql.Driver (PerfLogger.java:PerfLogBegin(100))
- <PERFLOG method=releaseLocks>
2013-11-15 11:56:49,874 INFO  ql.Driver (PerfLogger.java:PerfLogEnd(127))
- </PERFLOG method=releaseLocks start=1384516609874 end=1384516609874 duration=0>
```

The highlighted entries in Listing 13-8 are the regions you should be looking at if you wish to see the chain of events while executing your CREATE DATABASE Hive job.

Other entries are helpful in the event of an error. Say, for example, you try to create a database that already exists. The attempt would fail. You would then look for entries in the log file such as those highlighted in Listing 13-9.

Listing 13-9. hive.log file showing HQL errors

```
2013-11-15 13:37:11,432 INFO  ql.Driver (PerfLogger.java:PerfLogBegin(100))
- <PERFLOG method=Driver.run>
2013-11-15 13:37:11,433 INFO  ql.Driver (PerfLogger.java:PerfLogBegin(100))
- <PERFLOG method=TimeToSubmit>
2013-11-15 13:37:11,433 INFO  ql.Driver (PerfLogger.java:PerfLogBegin(100))
- <PERFLOG method=compile>
2013-11-15 13:37:11,434 INFO  parse.ParseDriver (ParseDriver.java:parse(179))
- Parsing command: create database test
2013-11-15 13:37:11,434 INFO  parse.ParseDriver (ParseDriver.java:parse(197))
- Parse Completed
2013-11-15 13:37:11,435 INFO  ql.Driver (Driver.java:compile(442))
- Semantic Analysis Completed
2013-11-15 13:37:11,436 INFO  ql.Driver (Driver.java:getSchema(259))
- Returning Hive schema: Schema(fieldSchemas:null, properties:null)
2013-11-15 13:37:11,436 INFO  ql.Driver (PerfLogger.java:PerfLogEnd(127))
- </PERFLOG method=compile start=1384522631433 end=1384522631436 duration=3>
2013-11-15 13:37:11,437 INFO  ql.Driver (PerfLogger.java:PerfLogBegin(100))
- <PERFLOG method=Driver.execute>
2013-11-15 13:37:11,437 INFO  ql.Driver (Driver.java:execute(1066))
- Starting command: create database test
2013-11-15 13:37:11,437 INFO  ql.Driver (PerfLogger.java:PerfLogEnd(127))
- </PERFLOG method=TimeToSubmit start=1384522631433 end=1384522631437 duration=4>
2013-11-15 13:37:11,508 ERROR exec.Task (SessionState.java:printError(432))
- Database test already exists
2013-11-15 13:37:11,509 ERROR ql.Driver (SessionState.java:printError(432))
- FAILED: Execution Error, return code 1 from org.apache.hadoop.hive.ql.exec.DDLTask
2013-11-15 13:37:11,510 INFO  ql.Driver (PerfLogger.java:PerfLogEnd(127))
- </PERFLOG method=Driver.execute start=1384522631437 end=1384522631510 duration=73>
2013-11-15 13:37:11,511 INFO  ql.Driver (PerfLogger.java:PerfLogBegin(100))
- <PERFLOG method=releaseLocks>
2013-11-15 13:37:11,512 INFO  ql.Driver (PerfLogger.java:PerfLogEnd(127))
- </PERFLOG method=releaseLocks start=1384522631511 end=1384522631512 duration=1>
2013-11-15 13:37:11,512 INFO  ql.Driver (PerfLogger.java:PerfLogBegin(100))
- <PERFLOG method=releaseLocks>
2013-11-15 13:37:11,514 INFO  ql.Driver (PerfLogger.java:PerfLogEnd(127))
- </PERFLOG method=releaseLocks start=1384522631512 end=1384522631513 duration=1>
```

Much the same way, if you try to drop a database that does not even exist, you would see errors logged like those in Listing 13-10.

Listing 13-10. hive.log file showing some errors

```
2013-11-15 14:25:31,810 INFO  ql.Driver (PerfLogger.java:PerfLogBegin(100))
- <PERFLOG method=Driver.run>
2013-11-15 14:25:31,811 INFO  ql.Driver (PerfLogger.java:PerfLogBegin(100))
- <PERFLOG method=TimeToSubmit>
2013-11-15 14:25:31,811 INFO  ql.Driver (PerfLogger.java:PerfLogBegin(100))
- <PERFLOG method=compile>
2013-11-15 14:25:31,812 INFO  parse.ParseDriver (ParseDriver.java:parse(179))
- Parsing command: drop database hive
2013-11-15 14:25:31,813 INFO  parse.ParseDriver (ParseDriver.java:parse(197))
- Parse Completed
2013-11-15 14:25:31,814 INFO  ql.Driver (Driver.java:compile(442))
- Semantic Analysis Completed
2013-11-15 14:25:31,815 INFO  ql.Driver (Driver.java:getSchema(259))
- Returning Hive schema: Schema(fieldSchemas:null, properties:null)
2013-11-15 14:25:31,816 INFO  ql.Driver (PerfLogger.java:PerfLogEnd(127))
- </PERFLOG method=compile start=1384525531811 end=1384525531816 duration=5>
2013-11-15 14:25:31,816 INFO  ql.Driver (PerfLogger.java:PerfLogBegin(100))
- <PERFLOG method=Driver.execute>
2013-11-15 14:25:31,816 INFO  ql.Driver (Driver.java:execute(1066))
- Starting command: drop database hive
2013-11-15 14:25:31,816 INFO  ql.Driver (PerfLogger.java:PerfLogEnd(127))
- </PERFLOG method=TimeToSubmit start=1384525531811 end=1384525531816 duration=5>
2013-11-15 14:25:31,846 ERROR exec.Task (SessionState.java:printError(432))
- There is no database named hive
NoSuchObjectException(message:There is no database named
hive)Atorg.apache.hadoop.hive.metastore.api.ThriftHiveMetastore
$get_database_result$get_database_resultStandardScheme.read(ThriftHiveMetastore.java:9883)
```

There could be errors while executing DML commands like SELECT against your Hive tables. To understand and troubleshoot such errors, you need to know the different phases that a HQL goes through. Table 13-1 summarizes the phases of Hive query execution.

Table 13-1. *Hive query execution phases*

Phases	Description
Parsing	Converts a Query into Parse Tree. If there are syntax errors in your query (for example, a missing semi-colon at the end), it is likely to be failing at this stage.
Semantic Analysis	Builds a logical plan based on the information retrieved from the Hive metastore database. Metadata failure errors, where the underlying schema has changed after the query is submitted, are reported in this phase.
Physical Plan Generation	Converts the logical plan to a physical plan that generates a Direct Acyclic Graph of MapReduce jobs that need to be executed. The errors reported in this stage or after this are MapReduce job errors. Further insights can be gained about these failures from the TaskTracker log files in the compute nodes.

Hive SELECT commands with aggregate functions (count, sum, and so on), or having conditions with column filters, invoke MapReduce jobs to get the command output. For example, if you execute the query SELECT count(*) from hivesampletable, you would see output with MapReduce job details as shown in Listing 13-11.

Listing 13-11. MapReduce Operation Log

```
Total MapReduce jobs = 1
Launching Job 1 out of 1
Number of reduce tasks determined at compile time: 1
In order to change the average load for a reducer (in bytes):
  set hive.exec.reducers.bytes.per.reducer=<number>
In order to limit the maximum number of reducers:
  set hive.exec.reducers.max=<number>
In order to set a constant number of reducers:
  set mapred.reduce.tasks=<number>
Starting Job = job_201311120315_0003,
Tracking URL = http://jobtrackerhost:50030
/jobdetails.jsp?jobid=job_201311120315_0003
Kill Command = c:\apps\dist\hadoop-1.2.0.1.3.0.1-0302\bin\
hadoop.cmd job  -kill job_201311120315_0003
Hadoop job information for Stage-1: number of mappers: 1; number of reducers: 1
2013-11-16 17:28:38,336 Stage-1 map = 0%,  reduce = 0%
2013-11-16 17:28:42,354 Stage-1 map = 100%,  reduce = 0%, Cumulative CPU 3.093 sec
2013-11-16 17:28:43,363 Stage-1 map = 100%,  reduce = 0%, Cumulative CPU 3.093 sec
2013-11-16 17:28:44,376 Stage-1 map = 100%,  reduce = 0%, Cumulative CPU 3.093 sec
2013-11-16 17:28:45,388 Stage-1 map = 100%,  reduce = 0%, Cumulative CPU 3.093 sec
2013-11-16 17:28:46,395 Stage-1 map = 100%,  reduce = 0%, Cumulative CPU 3.093 sec
2013-11-16 17:28:47,401 Stage-1 map = 100%,  reduce = 0%, Cumulative CPU 3.093 sec
2013-11-16 17:28:48,409 Stage-1 map = 100%,  reduce = 0%, Cumulative CPU 3.093 sec
2013-11-16 17:28:49,416 Stage-1 map = 100%,  reduce = 0%, Cumulative CPU 3.093 sec
2013-11-16 17:28:50,423 Stage-1 map = 100%,  reduce = 33%, Cumulative CPU 3.093sec
2013-11-16 17:28:51,429 Stage-1 map = 100%,  reduce = 33%, Cumulative CPU 3.093sec
2013-11-16 17:28:52,445 Stage-1 map = 100%,  reduce = 100%, Cumulative CPU 5.514 sec
2013-11-16 17:28:53,453 Stage-1 map = 100%,  reduce = 100%, Cumulative CPU 5.514 sec
2013-11-16 17:28:54,462 Stage-1 map = 100%,  reduce = 100%, Cumulative CPU 5.514 sec
MapReduce Total cumulative CPU time: 5 seconds 514 msec
Ended Job = job_201311120315_0003
MapReduce Jobs Launched:
Job 0: Map: 1  Reduce: 1   Cumulative CPU: 5.514 sec
HDFS Read: 245 HDFS Write: 6 SUCCESS
Total MapReduce CPU Time Spent: 5 seconds 514 msec
OK
59793
Time taken: 48.899 seconds, Fetched: 1 row(s)
```

As we see from the preceding output, the job that is created is job_201311120315_0003. Now, take a look at the folder C:\apps\dist\hadoop-1.2.0.1.3.0.1-0302\logs\. In that folder, you should have a file named job_201311120315_0003_conf.xml. The content of that file gives information about all the environment variables and configuration details for that MapReduce job.

The TaskTracker logs come into play when the Hive queries are through the physical plan-generation phase and into the MapReduce phase. From that point forward, TaskTracker logs will have a detailed tracing of the operations performed. Note that the individual tasks are executed on the data nodes, hence the TaskTracker logs are available in the data nodes only.

The NameNode maintains the log files for the JobTracker service in the same `C:\apps\dist\hadoop-1.2.0.1.3.0.1-0302\logs\.` folder. The JobTracker service is responsible for determining the location of the data blocks, maintaining co-ordination with and monitoring the TaskTracker services running on different data nodes. The file name is `Hadoop-jobtracker-<node name>.log`. You can open the file, and its contents should be similar to Listing 13-12.

Listing 13-12. The JobTracker Log

```
2013-11-16 17:28:29,781 INFO org.apache.hadoop.mapred.JobTracker:
Initializing job_201311120315_0003
2013-11-16 17:28:29,781 INFO org.apache.hadoop.mapred.JobInProgress:
Initializing job_201311120315_0003
2013-11-16 17:28:29,952 INFO org.apache.hadoop.mapred.JobInProgress:
Input size for job job_201311120315_0003 = 5015508. Number of splits = 1
2013-11-16 17:28:29,953 INFO org.apache.hadoop.mapred.JobInProgress:
tip:task_201311120315_0003_m_000000 has split on node:/fd0/ud0/localhost
2013-11-16 17:28:29,953 INFO org.apache.hadoop.mapred.JobInProgress:
job_201311120315_0003 LOCALITY_WAIT_FACTOR=0.25
2013-11-16 17:28:29,953 INFO org.apache.hadoop.mapred.JobInProgress:
Job job_201311120315_0003 initialized successfully with 1 map tasks and 1 reduce tasks.
2013-11-16 17:28:29,966 INFO org.apache.hadoop.mapred.JobTracker:
Adding task (JOB_SETUP) 'attempt_201311120315_0003_m_000002_0' to tip
task_201311120315_0003_m_000002, for tracker 'tracker_workernode2:127.0.0.1/127.0.0.1:49175'
2013-11-16 17:28:37,865 INFO org.apache.hadoop.mapred.JobInProgress:
Task 'attempt_201311120315_0003_m_000002_0' has completed task_201311120315_0003_m_000002 successfully.
2013-11-16 17:28:37,869 INFO org.apache.hadoop.mapred.JobInProgress:
Choosing a non-local task task_201311120315_0003_m_000000
2013-11-16 17:28:37,870 INFO org.apache.hadoop.mapred.JobTracker: Adding task (MAP)
'attempt_201311120315_0003_m_000000_0' to tip task_201311120315_0003_m_000000,
for tracker 'tracker_workernode2:127.0.0.1/127.0.0.1:49175'
2013-11-16 17:28:39,710 INFO org.apache.hadoop.mapred.JobInitializationPoller:
Removing scheduled jobs from waiting queuejob_201311120315_0003
2013-11-16 17:28:42,118 INFO org.apache.hadoop.mapred.JobInProgress:
Task 'attempt_201311120315_0003_m_000000_0' has completed task_201311120315_0003_m_000000 successfully.
2013-11-16 17:28:42,151 INFO org.apache.hadoop.mapred.JobTracker:
Adding task (REDUCE) 'attempt_201311120315_0003_r_000000_0' to tip
task_201311120315_0003_r_000000, for tracker 'tracker_workernode2:127.0.0.1/127.0.0.1:49175'
2013-11-16 17:28:51,473 INFO org.apache.hadoop.mapred.JobInProgress:
Task 'attempt_201311120315_0003_r_000000_0' has completed task_201311120315_0003_r_000000 successfully.
2013-11-16 17:28:51,484 INFO org.apache.hadoop.mapred.JobTracker:
Adding task (JOB_CLEANUP) 'attempt_201311120315_0003_m_000001_0' to tip
task_201311120315_0003_m_000001, for tracker 'tracker_workernode2:127.0.0.1/127.0.0.1:49175'
2013-11-16 17:28:53,734 INFO org.apache.hadoop.mapred.JobInProgress:
Task 'attempt_201311120315_0003_m_000001_0' has completed task_201311120315_0003_m_000001 successfully.
2013-11-16 17:28:53,735 INFO org.apache.hadoop.mapred.JobInProgress:
Job job_201311120315_0003 has completed successfully.
```

```
2013-11-16 17:28:53,736 INFO org.apache.hadoop.mapred.JobInProgress$JobSummary:
jobId=job_201311120315_0003,submitTime=1384622907254,
launchTime=1384622909953,firstMapTaskLaunchTime=1384622917870,
firstReduceTaskLaunchTime=1384622922122,firstJobSetupTaskLaunchTime=1384622909966,
firstJobCleanupTaskLaunchTime=1384622931484,finishTime=1384622933735,numMaps=1,
numSlotsPerMap=1,numReduces=1,numSlotsPerReduce=1,user=amarpb,queue=default,
status=SUCCEEDED,mapSlotSeconds=8,reduceSlotsSeconds=9,clusterMapCapacity=16,
clusterReduceCapacity=8,jobName=select count(*) from hivesampletable(Stage-1)
2013-11-16 17:28:53,790 INFO org.apache.hadoop.mapred.JobQueuesManager:
Job job_201311120315_0003 submitted to queue default has completed
2013-11-16 17:28:53,791 INFO org.apache.hadoop.mapred.JobTracker:
Removing task 'attempt_201311120315_0003_m_000000_0'
2013-11-16 17:28:53,791 INFO org.apache.hadoop.mapred.JobTracker:
Removing task 'attempt_201311120315_0003_m_000001_0'
2013-11-16 17:28:53,791 INFO org.apache.hadoop.mapred.JobTracker:
Removing task 'attempt_201311120315_0003_m_000002_0'
2013-11-16 17:28:53,792 INFO org.apache.hadoop.mapred.JobTracker:
Removing task 'attempt_201311120315_0003_r_000000_0'
2013-11-16 17:28:53,815 INFO org.apache.hadoop.mapred.JobHistory:
Creating DONE subfolder at wasb://democlustercontainer@democluster.blob.core.windows.net/mapred/
history/done/version-1/jobtrackerhost_1384226104721_/2013/11/16/000000
2013-11-16 17:28:53,978 INFO org.apache.hadoop.mapred.JobHistory:
Moving file:/c:/apps/dist/hadoop-1.2.0.1.3.0.1-
0302/logs/history/job_201311120315_0003_1384622907254_desarkar_
select+count%28%20F%29+from+hivesampletable%28Stage-1%29_default_%20F
to wasb://testhdi@democluster.blob.core.windows.net/mapred/history/done/
version-1/jobtrackerhost_1384226104721_/2013/11/16/000000
2013-11-16 17:28:54,322 INFO org.apache.hadoop.mapred.JobHistory:
Moving file:/c:/apps/dist/hadoop-1.2.0.1.3.0.1-0302/logs/history/
job_201311120315_0003_conf.xml to wasb://democlustercontainer@democluster.blob.core.windows.net/mapred/
history/done/version-1/jobtrackerhost_1384226104721_/2013/11/16/000000
```

The JobTracker log files are pretty verbose. If you go through them carefully, you should be able to track down and resolve any errors in your Hive data-processing jobs.

Troubleshooting can be tricky however, if the problem is with job performance. If your Hive queries are joining multiple tables and their different partitions, the query response times can be quite long. In some cases, they will need manual tuning for optimum throughput. To that end, the following subsections provide some best practices leading toward better execution performance.

Compress Intermediate Files

A large volume of intermediate files are generated during the execution of MapReduce jobs. Analysis has shown that if these intermediate files are compressed, job execution performance tends to be better. You can execute the following SET commands to set compression parameters from the Hadoop command line:

```
set mapred.compress.map.output=true;
set
mapred.map.output.compression.codec=org.apache.hadoop.io.compress.GzipCodec;
set hive.exec.compress.intermediate=true
```

▨ **Note**   Currently, HDInsight supports `Gzip` and `BZ2` codecs.

Configure the Reducer Task Size

In majority of the MapReduce job-execution scenarios, after the map jobs are over, most of the nodes go idle with only a few nodes working for the reduce jobs to complete. To make reduce jobs finish fast, you can increase the number of reducers to match the number of nodes or the total number of processor cores. Following is the SET command you use to configure the number of reducers launched from a Hive job:

```
set mapred.reduce.tasks=<number>
```

Implement Map Joins

Map joins in Hive are particularly useful when a single, huge table needs to be joined with a very small table. The small table can be placed into memory, in a distributed cache, by using map joins. By doing that, you avoid a good deal of disk IO. The SET commands in Listing 13-13 enable Hive to perform map joins and cache the small table in memory.

Listing 13-13. Hive SET options

```
set hive.auto.convert.join=true;
set hive.mapjoin.smalltable.filesize=40000000;
```

Another important configuration is the `hive.mapjoin.smalltable.filesize` setting. By default, it is 25 MB, and if the smaller table exceeds this size, all of your original MapJoin tests revert back to common joins. In the preceding snippet, I have overridden the default setting and set it to 40 MB.

▨ **Note**   There are no reducers in map joins, because such a join can be completed during the map phase with a lot less data movement.

You can confirm that map joins are happening if you see the following:

- With a map join, there are no reducers because the join happens at the map level.

- From the command line, it'll report that a map join is being done because it is pushing a smaller table up to memory.

- And right at the end, there is a call out that it's converting the join into `MapJoin`.

The command-line output or the Hive logs will have snippets indicating that a map join has happened, as you can see in Listing 13-14.

Listing 13-14. hive.log file

```
2013-11-26 10:55:41 Starting to launch local task to process map join;
maximum memory = 932118528
2013-11-26 10:55:45 Processing rows: 200000 Hashtable size: 199999
Memory usage: 145227488 rate: 0.158
2013-11-26 10:55:47 Processing rows: 300000 Hashtable size: 299999
Memory usage: 183032536 rate: 0.188
```

```
2013-11-26 10:55:49 Processing rows: 330936 Hashtable size: 330936
Memory usage: 149795152 rate: 0.166
2013-11-26 10:55:49 Dump the hashtable into file: file:/tmp/msgbigdata/
hive_2013-11-26 _22-55-34_959_3143934780177488621/-local-10002/
HashTable-Stage-4/MapJoin-mapfile01-.hashtable
2013-11-26 10:55:56 Upload 1 File to: file:/tmp/msgbigdata/
hive_2013-11-26 _22-55-34_959_3143934780177488621/-local-10002/
HashTable-Stage-4/MapJoin-mapfile01-.hashtable File size: 39685647
2013-11-26  10:55:56 End of local task; Time Taken: 13.203 sec.
Execution completed successfully
Mapred Local Task Succeeded . Convert the Join into MapJoin
Mapred Local Task Succeeded . Convert the Join into MapJoin
Launching Job 2 out of 2
```

Hive is a common choice in the Hadoop world. SQL users take no time to get started with Hive, because the schema-based data structure is very familiar to them. Familiarity with SQL syntax also translates well into using Hive.

Pig Jobs

Pig is a set-based, data-transformation tool that works on top of Hadoop and cluster storage. Pig offers a command-line application for user input called Grunt, and the scripts are called Pig Latin. Pig can be run on the name-node host or client machine, and it can run jobs that read data from HDFS/WASB and compute data using the MapReduce framework. The biggest advantage, again, is to free the developer from writing complex MapReduce programs.

Configuration File

The configuration file for Pig is pig.properties, and it is found in the C:\apps\dist\pig-0.11.0.1.3.1.0-06\conf\ directory of the HDInsight name node. It contains several key parameters for controlling job submission and execution. Listing 13-15 highlights a few of them.

Listing 13-15. pig.properties file

```
#Verbose print all log messages to screen (default to print only INFO and above to screen)
verbose=true

#Exectype local|mapreduce, mapreduce is default
exectype=mapreduce
#The following two parameters are to help estimate the reducer number
pig.exec.reducers.bytes.per.reducer=1000000000
pig.exec.reducers.max=999
#Performance tuning properties
pig.cachedbag.memusage=0.2
pig.skewedjoin.reduce.memusagea=0.3
pig.exec.nocombiner=false
opt.multiquery=true
pig.tmpfilecompression=false
```

These properties help you control the number of mappers and reducers, and several other performance-tuning options dealing with the internal dataset joins and memory usage.

░ **Tip** A very important debugging trick is to use the `exectype` parameter in Pig. By default, it is set to `exectype=mapreduce`, which means you need access to your cluster and its storage to run your scripts. You can set this to `exectype=local` for debugging. To run the scripts in local mode, no Hadoop or HDFS installation is required. All files are installed and run from your local host and file system.

It is also possible to run Pig in Debug mode, which prints out additional messages in the console during job execution. Debug mode also provides higher logging levels that can help with isolation of a given problem. The following command starts the Pig console in Debug mode:

```
c:\apps\dist\pig-0.11.0.1.3.1.0-06\bin>pig.cmd Ddebug=DEBUG
```

For every Pig job, there is a job-configuration file that gets generated. The file is located at `C:\apps\dist\hadoop-1.2.0.1.3.1.0-06\logs\` directory and named as `job_jobId_conf.xml`.

Log Files

Pig does not have a log file directory of its own. Rather, it logs its operations in the `C:\apps\dist\hadoop-1.2.0.1.3.1.0-06\logs\` folder. The name of the log file is `pig_<random_number>.log`. This file records a *Pig Stack Trace* for every failure that happens during a Pig job execution. A sample excerpt of such a trace is shown in Listing 13-16.

Listing 13-16. Pig Stack Trace

```
========================================================================
Pig Stack Trace
---------------
ERROR 1000: Error during parsing. Encountered " <IDENTIFIER> "exit "" at line 4, column 1.
Was expecting one of: <EOF> <EOL> ...
org.apache.pig.tools.pigscript.parser.ParseException:
Encountered " <IDENTIFIER> "exit "" at line 4, column 1.
Was expecting one of:  <EOF>
========================================================================
```

It is important to understand that for each of these supporting projects the underlying execution framework is still MapReduce. Thus, if a job failure occurs at the MapReduce phase, the JobTracker logs are the place to investigate.

Explain Command

The `EXPLAIN` command in Pig shows the logical and physical plans of the MapReduce jobs triggered by your `Pig Latin` statements. Following is the Pig statement we executed in Chapter 6 to aggregate and sort the output messages from the `Sample.log` file. We'll use it as the basis for an example. Launch the Pig command shell from the `c:\apps\dist\pig-0.11.0.1.3.1.0-06\bin\` folder, and type in the lines of script one after another:

```
LOGS = LOAD 'wasb:///example/data/sample.log';
LEVELS = foreach LOGS generate REGEX_EXTRACT($0,'(TRACE|DEBUG|INFO|WARN|ERROR|FATAL)', 1)
   as LOGLEVEL;
FILTEREDLEVELS = FILTER LEVELS by LOGLEVEL is not null;
```

If you wish to display the Logical, Physical, and MapReduce execution plans for the `FILTEREDLEVELS` object, you can now issue the following command: `EXPLAIN FILTEREDLEVELS`. This command should produce output similar to that in Listing 13-17.

Listing 13-17. The Explain command

```
grunt> EXPLAIN FILTEREDLEVELS;
2013-11-22 18:30:55,721 [main] WARN  org.apache.pig.PigServer -
Encountered Warning IMPLICIT_CAST_TO_CHARARRAY 1 time(s).
2013-11-22 18:30:55,723 [main] WARN  org.apache.pig.PigServer -
Encountered Warning USING_OVERLOADED_FUNCTION 1 time(s).
#-----------------------------------------------
# New Logical Plan:
#-----------------------------------------------
FILTEREDLEVELS: (Name: LOStore Schema: LOGLEVEL#78:chararray)
|
|---FILTEREDLEVELS: (Name: LOFilter Schema: LOGLEVEL#78:chararray)
    |   |
    |   (Name: Not Type: boolean Uid: 80)
    |   |
    |   |---(Name: IsNull Type: boolean Uid: 79)
    |       |
    |       |---LOGLEVEL:(Name: Project Type: chararray Uid: 78 Input: 0 Column: 0)
    |
    |---LEVELS: (Name: LOForEach Schema: LOGLEVEL#78:chararray)
        |   |
        |   (Name: LOGenerate[false] Schema: LOGLEVEL#78:chararray)
        |   |   |
        |   |   (Name: UserFunc(org.apache.pig.builtin.REGEX_EXTRACT) Type: chararray Uid: 78)
        |   |   |
        |   |   |---(Name: Cast Type: chararray Uid: 74)
        |   |   |   |
        |   |   |   |---(Name: Project Type: bytearray Uid: 74 Input: 0 Column: (*))
        |   |   |
        |   |   |---(Name: Constant Type: chararray Uid: 76)
        |   |   |
        |   |   |---(Name: Constant Type: int Uid: 77)
        |   |
        |   |---(Name: LOInnerLoad[0] Schema: #74:bytearray)
        |
        |---LOGS: (Name: LOLoad Schema: null)RequiredFields:null


#-----------------------------------------------
# Physical Plan:
#-----------------------------------------------
FILTEREDLEVELS: Store(fakefile:org.apache.pig.builtin.PigStorage) - scope-57
|
|---FILTEREDLEVELS: Filter[bag] - scope-53
    |   |
    |   Not[boolean] - scope-56
    |   |
    |   |---POIsNull[boolean] - scope-55
```

```
|        |
|        |---Project[chararray][0] - scope-54
|
|---LEVELS: New For Each(false)[bag] - scope-52
     |    |
     |    POUserFunc(org.apache.pig.builtin.REGEX_EXTRACT)[chararray] - scope-50
     |    |
     |    |---Cast[chararray] - scope-47
     |    |   |
     |    |   |---Project[bytearray][0] - scope-46
     |    |
     |    |---Constant((TRACE|DEBUG|INFO|WARN|ERROR|TOTAL)) - scope-48
     |    |
     |    |---Constant(1) - scope-49
     |
     |---LOGS:
```
Load(wasb://democlustercontainer@democluster.blob.core.windows.net/
sample.log:org.apache.pig.builtin.PigStorage) - scope-45

```
#------------------------------------------------
# Map Reduce Plan
#------------------------------------------------
MapReduce node scope-58
Map Plan
FILTEREDLEVELS: Store(fakefile:org.apache.pig.builtin.PigStorage) - scope-57
|
|---FILTEREDLEVELS: Filter[bag] - scope-53
     |    |
     |    Not[boolean] - scope-56
     |    |
     |    |---POIsNull[boolean] - scope-55
     |    |    |
     |    |    |---Project[chararray][0] - scope-54
     |
     |---LEVELS: New For Each(false)[bag] - scope-52
          |    |
          |    POUserFunc(org.apache.pig.builtin.REGEX_EXTRACT)[chararray] -scope-50
          |    |
          |    |---Cast[chararray] - scope-47
          |    |   |
          |    |   |---Project[bytearray][0] - scope-46
          |    |
          |    |---Constant((TRACE|DEBUG|INFO|WARN|ERROR|TOTAL)) - scope-48
          |    |
          |    |---Constant(1) - scope-49
          |
          |---LOGS:
```
Load(wasb://democlustercontainer@democluster.blob.core.windows.net/
sample.log:org.apache.pig.builtin.PigStorage) - scope-45--------
Global sort: false

The EXPLAIN operator's output is segmented into three sections:

- **Logical Plan**   The Logical Plan gives you the chain of operators used to build the relations, along with data type validation. Any filters (like NULL checking) that might have been applied early on also apply here.

- **Physical Plan**   The Physical Plan shows how the logical operators are actually translated as physical operators with some memory-optimization techniques that might have been used.

- **MapReduce Plan**   The MapReduce Plan shows how the physical operators are grouped into MapReduce jobs that would actually work on the cluster's data.

Illustrate Command

The ILLUSTRATE command is one of the best ways to debug Pig scripts. The command attempts to provide a reader-friendly representation of the data. ILLUSTRATE works by taking a sample of the output data and running it through the Pig script. But as the ILLUSTRATE command encounters operators that remove data (such as filter, join, etc.), it makes sure that some records pass through the operator and some do not. When necessary, it will manufacture records that look similar to the data set. For example, if you have a variable B, formed by grouping another variable A, the ILLUSTRATE command on variable B will show you the details of the underlying composite types. Type in the following command in the Pig shell to check this out:

```
A = LOAD 'data' AS (f1:int, f2:int, f3:int);
B = GROUP A BY (f1,f2);
ILLUSTRATE B;
```

This will give you output similar to what is shown here:

```
-------------------------------------------------------------------
| b  |group: tuple({f1: int,f2: int})|a: bag({f1: int,f2: int,f3: int})|
-------------------------------------------------------------------
|    | (8, 3)                        | {(8, 3, 4), (8, 3, 4)} |
-------------------------------------------------------------------
```

You can use the ILLUSTRATE command to examine the structure of relation or variable B. Relation B has two fields. The first field is named group and is of type tuple. The second field is name a, after relation A, and is of type bag.

■ **Note**   A variable is also called a *relation* in Pig Latin terms.

Sqoop Jobs

Sqoop is the bi-directional data-transfer tool between HDFS (again, WASB in Azure HDInsight service) and relational databases. In an HDInsight context, Sqoop is primarily used to import and export data to and from SQL Azure databases and the cluster storage. When you run a Sqoop command, Sqoop in turn runs a MapReduce task in the Hadoop Cluster (map only, and no reduce task). There is no separate log file specific to Sqoop. So you need to troubleshoot a Sqoop failure or performance issue pretty much the same way as a MapReduce failure or performance issue.

Windows Azure Storage Blob

The underlying storage infrastructure for Azure is known as Windows Azure Blob Storage (WABS). Microsoft has implemented a thin wrapper that exposes this blob storage as the HDFS file system for HDInsight. This is referred to as Windows Azure Storage Blob (WASB) and is a notable change in Microsoft's Hadoop implementation on Windows Azure.

As you saw throughout the book, *Windows Azure Storage Blob (WASB)* replaces HDFS and is the storage for your HDInsight clusters, by default. It is important to understand the WASB issues you may encounter during your job submissions because all your input files are in WASB, and all the output files written by Hadoop are also in your cluster's dedicated WASB container.

WASB Authentication

One of the most common errors encountered during cluster operations is the following:

```
org.apache.hadoop.fs.azure.AzureException:
Unable to access container <container> in account <storage_account>
using anonymous credentials, and no credentials found for them in the configuration.
```

This message essentially means that the WASB code couldn't find the key for the storage account in the configuration.

Typically, the problem is one of two things:

- The key is not present in `core-site.xml`. Or it is there, but not in the correct format. This is usually easy to check (assuming you can use Remote Desktop to connect to your cluster). Take a look in the cluster (in `C:\apps\dist\hadoop-1.2.0.1.3.1.0-06\conf\core-site.xml`) for the configuration name-value pair with the name being `fs.azure.account.key.<account>`.

- The key is there in `core-site.xml`, but the process running into this exception is not reading `core-site.xml`. Most Hadoop components (MapReduce, Hive, and so on) read `core-site.xml` from that location for their configuration, but some don't. For example, Oozie has its own copy of `core-site.xml` that it uses. This is harder to chase, but if you're using a non-standard Hadoop component, this might be the culprit.

You should confirm your storage account key from your Azure Management portal and make sure that you have the correct entry in the `core-site.xml` file.

Azure Throttling

Windows Azure Blob Storage limits the bandwidth per storage account to maintain high storage availability for all customers. Limiting bandwidth is done by rejecting requests to storage (HTTP response 500 or 503) in proportion to recent requests that are above the allocated bandwidth. To learn about such storage account limits, refer to the following page:

```
http://blogs.msdn.com/b/windowsazure/archive/2012/11/02/windows-azure-s-flat-network-
storage-and-2012-scalability-targets.aspx.
```

Your cluster will be throttled if or when your cluster is writing data to or reading data from WASB at rates greater than those stated earlier. You can determine if you might hit those limits based on the size of your cluster and your workload type.

■ **Note** Real Hadoop jobs have recurring task startup delays, so the actual number of machines required to exceed the limit is generally higher than calculated.

Some initial indications that your job is being throttled by Windows Azure Storage may include the following:

- Longer-than-expected job completion times
- A high number of task failures
- Job failure

Although these are indications that your cluster is being throttled, the best way to understand if your workload is being throttled is by inspecting responses returned by Windows Azure Storage. Responses with an http status code of 500 or 503 indicate that a request has been throttled. One way to collect Windows Azure Storage responses is to turn on storage logging as described in http://www.windowsazure.com/en-us/manage/services/storage/how-to-monitor-a-storage-account/#configurelogging. This is also discussed earlier in this book in Chapter 11.

To avoid throttling, you can adjust parameters in the WASB driver self-throttling mechanism. The WASB driver is the HDInsight component that reads data from and writes data to WASB. The driver has a self-throttling mechanism that can slow individual virtual machine (VM) transfer rates between a cluster and WASB. This effectively slows the overall transfer rate between a cluster and WASB. The rate at which the self-throttling mechanism slows the transfer rate can be adjusted to keep transfer rates below throttling thresholds.

By default, the self-throttling mechanism is exercised for clusters with n (number of nodes) >= 7, and it increasingly slows transfer rates as n increases. The default rate at which self-throttling is imposed is set at cluster creation time (based on the cluster size), but it is configurable after cluster creation.

The self-throttling algorithm works by delaying a request to WASB in proportion to the end-to-end latency of the previous request. The exact proportion is determined by the following parameters (configurable in core-site.xml or at job submission time):

```
fs.azure.selfthrottling.read.factor (used when reading data from WASB)
fs.azure.selfthrottling.write.factor (used when writing data to WASB)
```

■ **Note** Valid values for these settings are in the following range: (0, 1).

Example 1: If your cluster has *n=20* nodes and is primarily doing heavy write operations, you can calculate the appropriate fs.azure.selftthrottling.write.factor value (for a storage account with geo-replication on):

```
fs.azure.selfthrottling.write.factor = 5Gbps/(800Mbps * 20) = 0.32
```

Example 2: If your cluster has *n=20* nodes and is doing heavy read operations, you can calculate the appropriate fs.azure.selfthrottling.read.factor value (for a storage account with geo-replication off):

```
fs.azure.selfthrottling.read.factor = 15Gbps/(1600Mbps * 20) = 0.48
```

If you still find that throttling continues after adjusting the parameter values just shown, further analysis and adjustment may be necessary.

Connectivity Failures

There are a few ways you can connect to your cluster. You can use remote desktop login to connect to the head node, you can use the ODBC endpoint on port *443* to connect to the Hive service, and you can navigate through the REST-based protocols to different URLs from Internet Explorer.

Always, make sure to test these different types of connections when you encounter a specific problem. For example, if you are unable to remotely log in to one of your data nodes, try to open the Hadoop Name Node Status portal and check if the number of live nodes is reported correctly.

■ **Note** Azure VMs periodically go through a process called *re-imaging*, where an existing VM is released and a new VM gets provisioned. The node is expected to be down for up to 15 minutes when this happens. This is an unattended, automated process, and the end user has absolutely no control over this.

ODBC failures deserve some additional attention. You typically use a client like Microsoft Excel to create your data models from HDInsight data. Any such front-end tool leverages the Hive ODBC driver to connect to Hive running on HDInsight. A typical failure can be like this:

Errors:
From Excel:
"Unable to establish connection with hive server"
From PowerPivot:
Failed connect to the server. Reason: ERROR [HY000] Invalid Attribute Persist Security Info
ERROR [01004] Out connection string buffer not allocated
ERROR [08001] Unable to establish connection with hive server

To start with, always make sure that the basic DSN configuration parameters such as port number, authentication, and so on are properly set. For Azure HDInsight Service, make sure that:

- You are connecting to port *443*.

- Hive Server Type is set to *Hive Server 2*.

- Authentication Mechanism is set to *Windows Azure HDInsight Service*.

- The correct cluster user name and password are provided.

For the Azure HDInsight Emulator, confirm that

- You are connecting to port *10001*.

- Hive Server Type is set to *Hive Server 2*.

- Authentication Mechanism is set to *Windows Azure HDInsight Emulator*.

If the problem persists even when all the preceding items are set correctly, try to test basic connectivity from Internet Explorer. Navigate to the following URLs, which target the same endpoints that ODBC uses:

Azure: https://<cluster>.azurehdinsight.net:443/hive/servlets/thrifths

Localhost: http://localhost:10001/servlets/thrifths

A successful test will show an HTTP 500 error where the error page will look like this at the top:

```
HTTP ERROR 500
Problem accessing /servlets/thrifths. Reason:
    INTERNAL_SERVER_ERROR
```

This error occurs because the server expects a specific payload to be sent in a request, and Internet Explorer doesn't allow for you to do that. However, the error does mean that the server is running and listening on the right port, and in that sense this particular error is actually a success.

For more help, you can turn on ODBC logging as described in Chapter 11. With logging on, you can trace each of the ODBC Driver Manager calls to investigate whatever problem is occurring.

Summary

The entire concept of using Azure HDInsight Service is based on the fact it is an elastic service—that is, a service you can extend as and when required. Submitting jobs is the only time you really need to spin up a cluster, because your data is always with you, residing on the Windows Azure Storage Blob, independent of your cluster.

It is very important to react and take corrective actions quickly when there is a job failure. This chapter focused on different types of jobs you can submit to your cluster and how to troubleshoot such a job failure. The chapter also covered some of the key Azure storage-related settings that could come in handy while troubleshooting an error or a performance problem, as well as the steps to diagnose connectivity failures to your cluster using the Hive ODBC driver.

Index

Get the eBook for only $10!

Now you can take the weightless companion with you anywhere, anytime. Your purchase of this book entitles you to 3 electronic versions for only $10.

This Apress title will prove so indispensible that you'll want to carry it with you everywhere, which is why we are offering the eBook in 3 formats for only $10 if you have already purchased the print book.

Convenient and fully searchable, the PDF version enables you to easily find and copy code—or perform examples by quickly toggling between instructions and applications. The MOBI format is ideal for your Kindle, while the ePUB can be utilized on a variety of mobile devices.

Go to www.apress.com/promo/tendollars to purchase your companion eBook.

CPSIA information can be obtained at www.ICGtesting.com
Printed in the USA
BVOW03s1732080614

355758BV00002BA/10/P